Information Circular 9521

Best Practices for Dust Control in Metal/Nonmetal Mining

By Jay F. Colinet, Andrew B. Cecala, Gregory J. Chekan, John A. Organiscak, and Anita L. Wolfe

DEPARTMENT OF HEALTH AND HUMAN SERVICES
Centers for Disease Control and Prevention
National Institute for Occupational Safety and Health
Office of Mine Safety and Health Research
Pittsburgh, PA • Spokane, WA

May 2010

This document is in the public domain and may be freely copied or reprinted.

Disclaimer

Mention of any company or product does not constitute endorsement by the National Institute for Occupational Safety and Health (NIOSH). In addition, citations to Web sites external to NIOSH do not constitute NIOSH endorsement of the sponsoring organizations or their programs or products. Furthermore, NIOSH is not responsible for the content of these Web sites. All Web addresses referenced in this document were accessible as of the publication date.

Ordering Information

To receive documents or other information about occupational safety and health topics, contact NIOSH at

>   Telephone: **1–800–CDC–INFO** (1–800–232–4636)
>   TTY: 1–888–232–6348
>   e-mail: cdcinfo@cdc.gov
>
>   or visit the NIOSH Web site at **www.cdc.gov/niosh**.

For a monthly update on news at NIOSH, subscribe to NIOSH *eNews* by visiting **www.cdc.gov/niosh/eNews**.

DHHS (NIOSH) Publication No. 2010–132

May 2010

SAFER • HEALTHIER • PEOPLE™

CONTENTS

Page

Introduction 1
Chapter 1.—Health effects of overexposure to respirable silica dust 3
    Silicosis 3
    Diagnosis and treatment 5
    References 6
Chapter 2.—Sampling to quantify respirable dust generation 8
    Respirable dust samplers for use in mining 8
    Sampling strategies 10
    References 13
Chapter 3.—Controlling respirable silica dust in underground stone and metal/nonmetal mines 15
    Crushing facilities 15
    Production shots 19
    Mucking operations 21
    Drilling 23
    References 23
Chapter 4.—Controlling respirable silica dust in mineral processing operations 26
    Primary dumping 27
    Crushing and grinding 30
    Transfer points 31
    Conveying 32
    Wet suppression 34
    Local exhaust ventilation (LEV) systems 38
    Low velocity transport systems 39
    Total mill ventilation systems 41
    Operator booths, control rooms, enclosed cabs 43
    Screening 45
    Packaging/bagging product for shipment 46
    Clothes cleaning system 50
    Background issues 54
    References 58
Chapter 5.—Controlling respirable silica dust at surface mines 64
    Drill dust collection systems 65
    Enclosed cab filtration systems 67
    Controlling haulage road dust 69
    Controlling dust at the primary hopper dump 70
    References 72

ILLUSTRATIONS

1-1. Section of freeze-dried human lung with silicosis 4
1-2. Inspector samples for select occupations from 2004-2008 that exceeded the PEL 5
2-1. Gravimetric sampling pump, cyclone and filter cassette 8
2-2. Example of dust measurements obtained with pDR 9

CONTENTS—Continued

Page

2-3. Personal dust monitor (PDM) with TEOM unit removed and shown on right 10
2-4. Sampling locations used to isolate dust generated at an underground crusher 11
2-5. Dust samplers mounted on haul truck in gold mine 12
2-6. Average dust generated by each segment of sampled haul truck cycle 12
2-7. Sampling locations around surface drill 13
3-1. Typical method used to isolate crushing facility from mine air 16
3-2. Closed ventilation system using plenum 17
3-3. Filtration and pressurization system components and design 18
3-4. Canopy air curtain blows filtered air over worker 19
3-5. Movement of dust after production shot as recorded by real-time dust sampler 20
3-6. Axial vane fan (left) versus propeller fan (right) 21
3-7. Fans positioned to ventilate dead-end entries 22
4-1. Staging curtains reduce dust billowing during dumping 29
4-2. Plastic stripping holds dust inside enclosure allowing water sprays to knock down dust ... 29
4-3. Tire-stop water spray system reduces rollback under dumping mechanism 30
4-4. Techniques for reducing respirable dust liberation from conveyor belts and transfers 34
4-5. Spray nozzles commonly used in mineral processing operations 36
4-6. Airborne dust capture performance of four types of spray nozzles 37
4-7. Canister-type dust collector system 39
4-8. Sawtooth design for low velocity transport system 41
4-9. Design concept of TMVS showing clean-air intakes and dust-laden air exhausts 42
4-10. Airflow pattern for one-directional filtration system for an enclosed cab 44
4-11. Screening unit with LEV system 45
4-12. Dual bag nozzle design 46
4-13. Bag and belt cleaner device 47
4-14. Semiautomated pallet loading system using push-pull ventilation 49
4-15. Telescoping bulk loading spout with an exhaust system 50
4-16. Clothes cleaning system design 52
4-17. Test subject before and after using the clothes cleaning booth 53
4-18. Drawing of a conventional structure and an open structure with a protective overhang .. 56
4-19. Overhead air supply island system 57
5-1. Typical dry dust collection system used on surface drills 65
5-2. Water separator discharging water before it reaches the drill bit 67
5-3. Increase in dust when a haul road dries 70
5-4. Staging curtains used to prevent dust from billowing out of enclosure 71
5-5. Tire-stop water spray system reduces dust rollback under the dumping vehicle 72

TABLES

4-1. Percent of samples exceeding PEL for select occupations 26
5-1. Respirable dust sampling results of enclosed cab field studies 68

ACRONYMS AND ABBREVIATIONS USED IN THIS REPORT

ACPH	air changes per hour
ACGIH	American Conference of Governmental Industrial Hygienists
CT	computed tomography
HEPA	high efficiency particulate air
HVAC	heating, ventilation, and air conditioning
IARC	International Agency for Research on Cancer
LEV	local exhaust ventilation
LHD	load-haul-dump
MSHA	Mine Safety and Health Administration
NIOSH	National Institute for Occupational Safety and Health
OASIS	overhead air supply island system
PDM	personal dust monitor
pDR	personal DataRAM
PEL	permissible exposure limit
PMF	progressive massive fibrosis
PPE	personal protective equipment
PVC	poly vinyl chloride
TEOM	tapered-element oscillating microbalance
TMVS	total mill ventilation system
XRD	X-ray diffraction

UNIT OF MEASURE ABBREVIATIONS USED IN THIS REPORT

cfm	cubic foot per minute
fpm	foot per minute
gpm	gallon per minute
in w.g.	inches water gauge
lpm	liter per minute
mg/m^3	milligram per cubic meter
mm	millimeter
mph	miles per hour
$\mu g/m^3$	microgram per cubic meter
psi	pound-force per square inch

BEST PRACTICES FOR DUST CONTROL IN METAL/NONMETAL MINING

By Jay F. Colinet,[1] Andrew B. Cecala,[2] Gregory J. Chekan,[2] John A. Organiscak,[2] and Anita L. Wolfe[3]

INTRODUCTION

Respirable silica dust exposure has long been known to be a serious health threat to workers in many industries. Overexposure to respirable silica dust can lead to the development of silicosis—a lung disease that can be disabling and fatal in its most severe form. Once contracted, there is no cure for silicosis so the goal must be to prevent development by limiting a worker's exposure to respirable silica dust. In addition, the International Agency for Research on Cancer (IARC) has concluded that there is sufficient evidence to classify silica as a human carcinogen.

For workers in the metal/nonmetal mining industry, the Mine Safety and Health Administration (MSHA) regulates and monitors exposure to respirable silica dust through personal dust sampling. Recent MSHA personal sampling results indicate that overexposures to respirable silica dust continue to occur for miners in metal/nonmetal mining operations. From 2004 to 2008, the percentages of samples that exceeded the applicable respirable dust standard for the different mining commodities were:

- 12% for sand and gravel
- 13% for stone
- 18% for nonmetal
- 21% for metal

Of the 2,407 deaths attributed to silicosis in the United States from 1990–1999, employment information was available for 881 deaths. Metal/nonmetal mining was the industry recorded for over 15% of these 881 deaths, with mining machine operator the most frequently recorded occupation.

In light of ongoing silica overexposures and reported silicosis deaths in metal/nonmetal miners, an ongoing threat to miners' health is evident. This handbook was developed to identify available engineering controls that can assist the industry in reducing worker exposure to respirable silica dust. The controls discussed in this handbook range from long-used controls which have developed into industry standards, to newer controls, which are still being optimized. The intent is to identify the "best practices" that are available for controlling respirable dust

---

[1]Senior scientist, Office of Mine Safety and Health Research, National Institute for Occupational Safety and Health, Pittsburgh, PA.
[2]Mining engineer, Office of Mine Safety and Health Research, National Institute for Occupational Safety and Health, Pittsburgh, PA.
[3]Public health advisor, Division of Respiratory Disease Studies, National Institute for Occupational Safety and Health, Morgantown, WV.

levels in underground and surface metal/nonmetal mining operations. This handbook provides general information on the control technologies along with extensive references. In some cases, the full reference(s) will need to be accessed to gain in-depth information on the testing or implementation of the control of interest.

The handbook is divided into five chapters. Chapter 1 discusses the health effects of exposure to respirable silica dust, while Chapter 2 discusses dust sampling instruments and sampling methods. Chapters 3, 4 and 5 are focused upon dust control technologies for underground mining, mineral processing, and surface mining, respectively.

Finally, it must be stressed that after control technologies are implemented, the ultimate success of ongoing protection for workers is dependent upon continued maintenance of these controls. On numerous occasions, National Institute for Occupational Safety and Health (NIOSH) researchers have seen appropriate controls installed, but worker overexposures continued to occur in the absence of proper maintenance of these controls.

CHAPTER 1. HEALTH EFFECTS OF OVEREXPOSURE TO RESPIRABLE SILICA DUST

By Anita Wolfe and Jay Colinet

Pneumoconioses are lung diseases caused by the inhalation and deposition of mineral dusts in the lungs. Known pneumoconioses include, but are not limited to, coal workers' pneumoconiosis and silicosis. These diseases are usually associated with working in a high-risk, mineral-related industry such as mining.

SILICOSIS

Occupational exposures to respirable crystalline silica occur in a variety of industries and occupations because of its extremely common natural occurrence. Respirable crystalline silica is defined as particles with aerodynamic diameters less than 10 microns [NIOSH 2002]. Workers with high exposure to crystalline silica include miners, sandblasters, tunnel builders, silica millers, quarry workers, foundry workers, and ceramics or glass workers. Silica refers to the chemical compound silicon dioxide (SiO_2), which occurs in a crystalline or noncrystalline (amorphous) form [NIOSH 2002]. Crystalline silica may be found in more than one form: alpha quartz, beta quartz, tridymite, and cristobalite [USBM 1992a; Heaney 1994]. In nature, the alpha form of quartz is the most common [Virta 1993]. This form is so abundant that the term quartz is often used in place of the general term crystalline silica [USBM 1992b; Virta 1993].

Quartz is a common component of rocks; consequently, mine workers are potentially exposed to quartz dust when rock is cut, drilled, crushed, and transported. Occupational exposures to respirable crystalline silica are associated with the development of silicosis, lung cancer, pulmonary tuberculosis, and airways diseases. These exposures may also be related to the development of autoimmune disorders, chronic renal disease (loss of kidney function), and other adverse health effects. In 1996 and 2009, the International Agency for Research on Cancer (IARC) reviewed the published experimental and epidemiologic studies of cancer in animals and workers exposed to respirable crystalline silica and concluded that there was sufficient evidence to classify silica as a human carcinogen [IARC 1997; Straif et al. 2009].

Silicosis is also a fibrosing disease of the lungs caused by the inhalation, retention, and pulmonary reaction to the crystalline silica. When silicosis becomes symptomatic, the primary symptom is usually dyspnea (difficult or labored breathing and/or shortness of breath), first noted with activity or exercise and later, as the functional reserve of the lung is also lost, at rest. However, in the absence of other respiratory diseases, there may be no shortness of breath and the disease may first be detected through an abnormal chest x-ray. The x-ray may at times show quite advanced disease with only minimal symptoms. The appearance or progression of dyspnea may indicate the development of complications including tuberculosis, airways obstruction, progressive massive fibrosis (PMF), or cor pulmonale (enlargement of the right side of the heart). Productive cough is often present.

A worker may develop one of three types of silicosis, depending on the airborne concentrations of respirable crystalline silica:

(1) *Chronic Silicosis:* Usually occurs after 10 or more years of exposure at relatively low concentrations. Swellings caused by the silica dust form in the lungs and lymph nodes of the chest. This disease may cause people to have trouble breathing and may be similar to chronic obstructive pulmonary disease.

(2) *Accelerated Silicosis:* Develops 5 to 10 years after the first exposure. Swelling in the lungs and symptoms occur faster than in chronic silicosis.

(3) *Acute Silicosis:* Develops after exposure to high concentrations of respirable crystalline silica and results in symptoms within a period of a few weeks to 5 years after the initial exposure [NIOSH 1986; Parker and Wagner 1998]. The lungs become very inflamed and can fill with fluid, causing severe shortness of breath and low blood oxygen levels.

PMF can occur in either simple or accelerated silicosis but is more common in the accelerated form. Figure 1-1 shows a lung that has been damaged by silicosis.

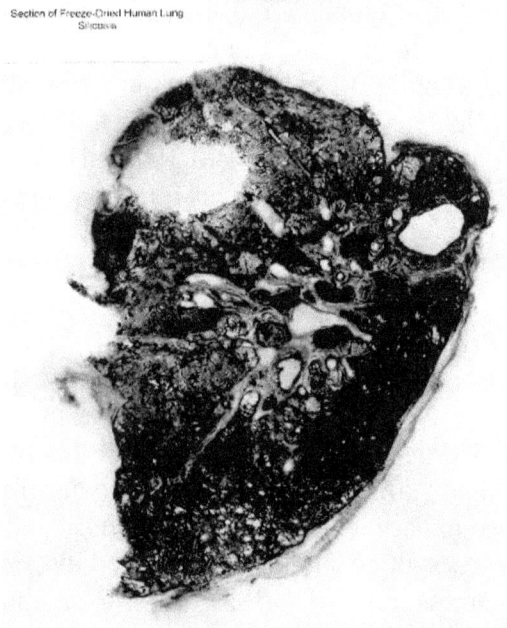

Figure 1-1. Section of freeze-dried human lung with silicosis.

In an effort to prevent the development of silicosis in miners working in metal/nonmetal mines, MSHA regulates their exposure to respirable silica. When quartz levels in respirable dust samples are greater than 1%, a respirable dust standard (permissible exposure limit) is calculated by dividing 10 mg/m^3 by the sum of the percent quartz plus 2. For example, if a sample contains 8% quartz, the respirable standard would be equal to 1 mg/m^3 (i.e., 10 ÷ (8 + 2)). This regulation places the upper limit of exposure to respirable quartz at 100 µg/m^3.

MSHA compliance sampling data identify those occupations in metal/nonmetal mining that are high-risk occupations for overexposure to quartz. Figure 1-2 shows the percent of samples collected by MSHA inspectors that exceeded the permissible exposure limit (PEL) for a number of high-risk occupations in metal/nonmetal mining.

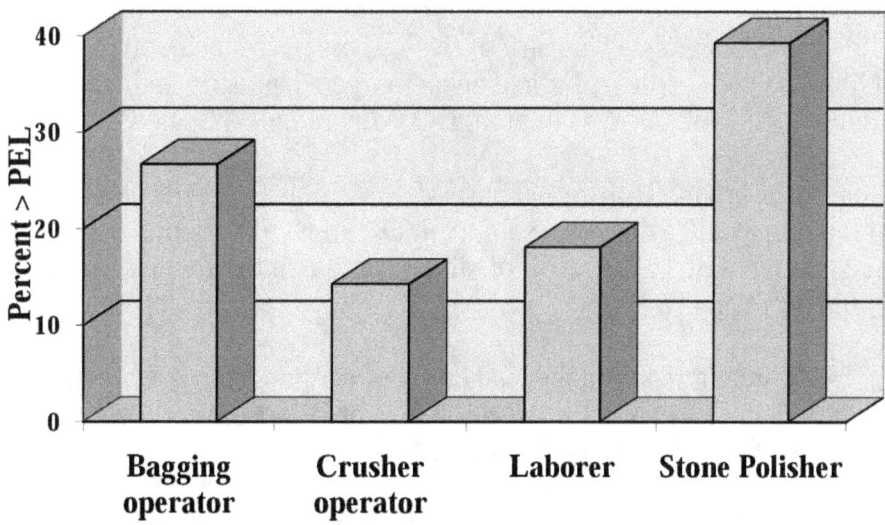

Figure 1-2. Inspector samples for select occupations from 2004–2008 that exceeded the PEL.

DIAGNOSIS AND TREATMENT

A doctor may diagnose silicosis based on the combination of an appropriate history of exposure to silica dust, compatible changes in chest imaging or lung pathology, and absence of plausible alternative diagnoses. A chest radiograph is often sufficient for diagnosis, but in some cases a computed tomography (CT) scan of the chest can be helpful. Lung biopsy, a procedure where a sample of lung tissue is taken for lab examination, is not usually required if a compatible exposure history and findings on chest imaging are present. Pulmonary function tests and blood tests to measure the amounts of oxygen and carbon dioxide in the blood (arterial blood gases) can help in objectively assessing the level of impairment caused by silicosis.

Epidemiologic studies of gold miners in South Africa, granite quarry workers in Hong Kong, metal miners in Colorado, and coal miners in Scotland have shown that chronic silicosis may develop or progress even after occupational exposure to silica has been discontinued [Hessel et al. 1988; Hnizdo and Sluis-Cremer 1993; Ng et al. 1987; Kreiss and Zhen 1996; Miller et al. 1998]. Therefore, removing a worker from exposure after diagnosis does not guarantee that silicosis or silica-related disease will stop progressing or that an impaired worker's condition will stabilize.

Treatment of silicosis may include use of bronchodilators (medications to open the airways) or supplemental oxygen. Once disease is detected, it is important to protect the lungs against respiratory infections, therefore a doctor may recommend vaccinations to prevent influenza and pneumonia. In some cases of severe disease, a lung transplant may be recommended. Prognosis depends on the length and level of exposure to respirable quartz dust. There is no cure for this lung disease and it cannot be reversed. Consequently, control technologies must be implemented in an effort to prevent the development of the disease. As an added measure of protection, a respirator program can be implemented for workers exposed to silica dust.

REFERENCES

Heaney PJ [1994]. Structure and chemistry of the low-pressure silica polymorphs. In: Heaney PJ, Prewitt CT, Gibbs GV, eds. Silica: physical behavior, geochemistry, and materials applications. Reviews in mineralogy. Vol. 29. Washington, DC: Mineralogical Society of America, pp.140.

Hessel PA, Sluis-Cremer GK, Hnizdo E, Faure MH, Thomas RG, Wiles FJ [1988]. Progression of silicosis in relation to silica dust exposure. Ann Occup Hyg 32(Suppl 1):689–696.
Hnizdo E, Sluis-Cremer GK [1993]. Risk of silicosis in a cohort of white South African gold miners. Am J Ind Med 24:447–457.

IARC [1997]. IARC monographs on the evaluation of carcinogenic risks to humans: silica, some silicates, coal dust and para-aramid fibrils. Vol 68. Lyon, France: World Health Organization, International Agency for Research on Cancer.

Kreiss K, Zhen B [1996]. Risk of silicosis in a Colorado mining community. Am J Ind Med 30:529–539.

Miller BG, Hagen S, Love RG, Soutar CA, Cowie HA, Kidd MW, Robertson A [1998]. Risks of silicosis in coal workers exposed to unusual concentrations of respirable quartz. Occup Environ Med 55:52–58.

Ng TP, Chan SL, Lam KP [1987]. Radiological progression and lung function in silicosis: a ten year follow up study. Br Med J 295:164–168.

NIOSH [1986]. Silicosis. By Peters JM. In: Merchant JA, Boehlecke BA, Taylor G, Pickett-Harner M, eds. Occupational respiratory diseases. Cincinnati, OH: U.S. Department of Health and Human Services, Centers for Disease Control, National Institute for Occupational Safety and Health, DHHS (NIOSH) Publication No. 86-102, pp. 219–237.

NIOSH [2002]. NIOSH hazard review: health effects of occupational exposure to respirable crystalline silica. Cincinnati, OH: U.S. Department of Health and Human Services, Centers for Disease Control and Prevention, National Institute for Occupational Safety and Health, DHHS (NIOSH) Publication No. 2002-129.

Parker JE, Wagner GR [1998]. Silicosis. In: Stellman JM, ed. Encyclopaedia of occupational health and safety. 4th ed. Geneva, Switzerland: International Labour Office, pp. 10.43–10.46.

Straif K, Benbrahim-Tallaa L, Baan R, Grosse Y, Secretan B, El Ghissassi F, Bouvard V, Guha N, Freeman C, Galichet L, Cogliano V [2009]. A review of human carcinogens—Part C: metals, arsenic, dusts, and fibres. Lancet Oncol 10(5):453–454.

USBM [1992a]. Crystalline silica overview: occurrence and analysis. By Ampian SG, Virta RL. Washington, DC: U.S. Department of the Interior, U.S. Bureau of Mines, Information Circular IC 9317.

USBM [1992b]. Crystalline silica primer. Washington, DC: U.S. Department of the Interior, U.S. Bureau of Mines.

Virta RL [1993]. Crystalline silica: what it is and isn't. Minerals Today Oct:12–16.

CHAPTER 2. SAMPLING TO QUANTIFY RESPIRABLE DUST GENERATION

By Jay F. Colinet

The respirable fraction of airborne dust is the dust that reaches the lungs and leads to the development of silicosis. Respirable dust cannot be seen with the eye. Conversely, if a dust cloud is visible, it is likely that a portion of the airborne dust will be in the respirable size range. In order to quantify the amount of harmful respirable dust that is in the mine air, sampling instrumentation must be used.

RESPIRABLE DUST SAMPLERS FOR USE IN MINING

The most common type of sampler used in the mining industry is the gravimetric sampler, which is designated by the Federal Coal Mine Health and Safety Act of 1969 for use in compliance dust sampling (Figure 2-1). This sampler consists of a constant-flow sampling pump, a size-selective cyclone, and a filter cartridge. In metal/nonmetal mining operations, the pump should be operated at 1.7 lpm. The 10-mm Dorr-Oliver cyclone separates the oversize dust from the respirable fraction (usually considered to have an aerodynamic diameter of 10 microns or less). The oversize dust is deposited into the grit pot at the bottom of the cyclone, while the respirable fraction is deposited onto a 37-mm-diameter polyvinyl chloride (PVC) filter. Care must be taken after a sample is collected to ensure that the cyclone assembly stays in an upright position. Otherwise, the oversize dust particles that are in the grit pot can be deposited onto the filter and invalidate the sample. The filter collects the respirable dust and is weighed to determine the mass of dust that has been collected during sampling. The mass of dust and the volume of sampled air are used to calculate the average concentration of respirable dust in mg/m^3.

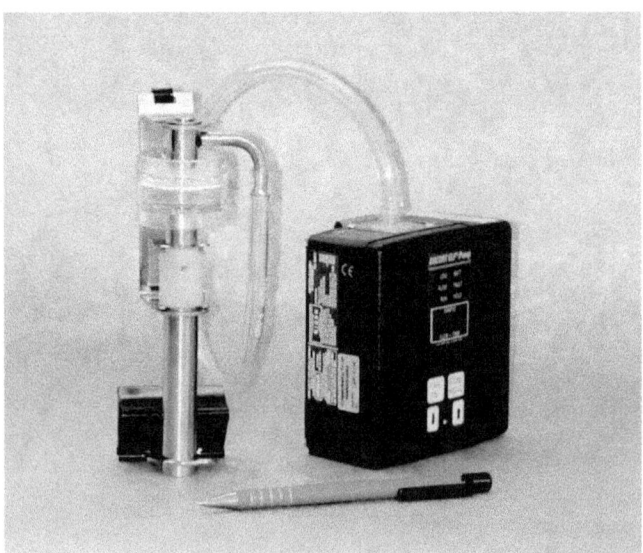

Figure 2-1. Gravimetric sampling pump, cyclone and filter cassette.

In order to determine the silica content of a gravimetric sample, the filter should be sent to an accredited laboratory for analysis. For samples collected in metal/nonmetal mines, x-ray diffraction (XRD) using NIOSH Analytical Method 7500 [NIOSH 1994, 2003] is the analytical technique used by MSHA's accredited laboratory [Parobeck and Tomb 2000] to quantify silica in the samples for compliance purposes.

In addition to the gravimetric samplers, a real-time dust sampler is available for use in mining. The personal DataRam (pDR) has dust-laden air pass through a sensing chamber in the sampler and passes a light beam through this dust. A sensor in the sampler measures the amount of light scatter caused by the dust and relates this scatter to a relative dust concentration. This concentration is correlated to the time when the sample was measured and stores this information in the internal data logger. The sample data can then be downloaded to a computer for analysis. Figure 2-2 illustrates a typical graph obtained with the pDR, as well as, a photo of the pDR. This data can be analyzed for specific time intervals (e.g., loading a cut), with average dust concentrations calculated for these intervals.

Unfortunately, the accuracy of the light-scattering monitor can be compromised by measuring dust clouds with different size distributions, different dust compositions, and/or water mist/fog in the ambient air. Consequently, when NIOSH utilizes pDR samplers, gravimetric samplers are placed adjacent to the pDR and individual pDR dust measurements are adjusted based upon the ratio between the average gravimetric concentration and the average pDR concentration [Thermo Scientific 2008]. For example, if the gravimetric concentration was 1.3 mg/m^3 over a 6-hour measurement period and the pDR average concentration was 1.0 mg/m^3 for the same 6 hours, then all individual pDR measurements would be multiplied by 1.3.

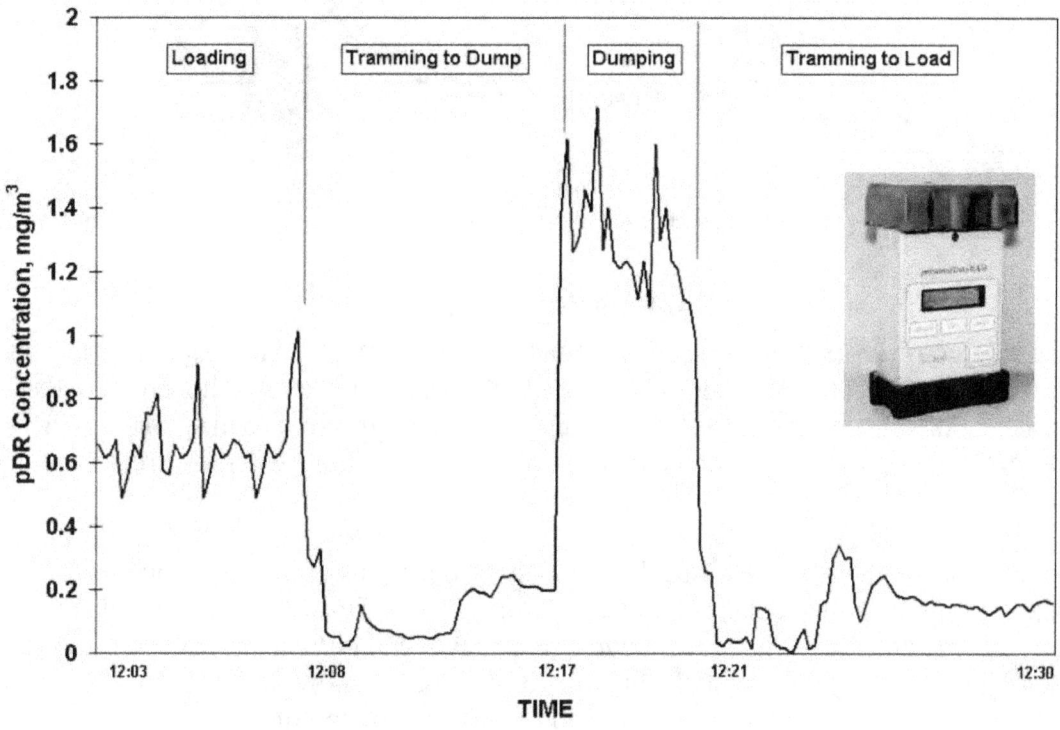

Figure 2-2. Example of dust measurements obtained with pDR.

The personal dust monitor (PDM) [NIOSH 2006] is another real-time sampler, which has been developed and tested by NIOSH, has been approved for use in underground mines by MSHA, and has reached commercial production. The PDM uses tapered element oscillating microbalance (TEOM) technology to obtain a real-time, gravimetric-based measure of respirable dust concentrations. The TEOM is a hollow tube that vibrates at a known frequency and has a filter mounted on the end. As respirable dust is deposited onto this filter, the TEOM frequency changes and this change can be related to a dust concentration. The PDM provides the wearer with a readout of the cumulative dust concentration to that point in the shift and the percent of the permissible exposure limit (PEL) that has been reached. This information can be used by the wearer to reduce their dust exposure prior to becoming overexposed. The sampler is incorporated into a standard cap lamp housing and has the sampling inlet located at the cap lamp (see Figure 2-3).

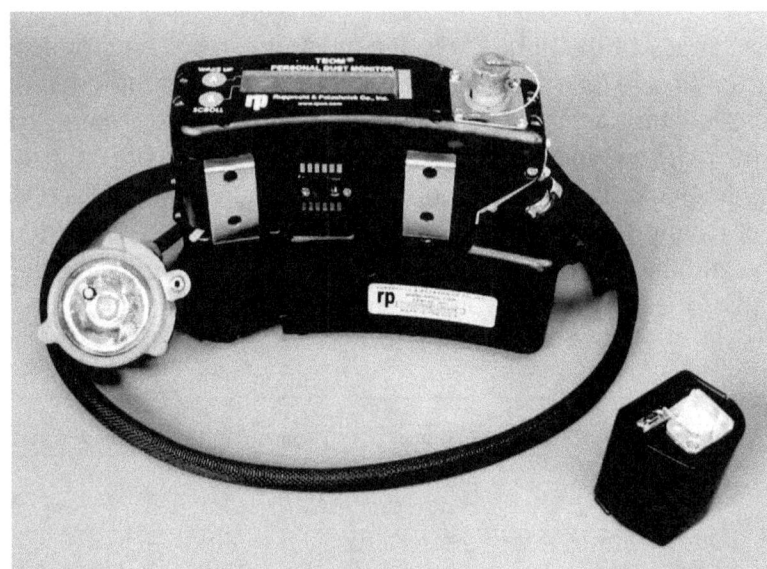

Figure 2-3. Personal dust monitor (PDM) with TEOM unit removed and shown on right.

SAMPLING STRATEGIES

In order to effectively control the respirable silica dust exposure of mine workers, it is necessary to identify the sources of dust generation and quantify the amount of dust liberated by these sources. Once the dust sources are identified and dust liberation from each source has been quantified, appropriate dust control technologies can be applied that offer the greatest protection to the mine workers.

In order to quantify the amount of dust liberated by a source, area dust sampling can be conducted in a manner that isolates the potential dust source. This is achieved by placing dust samplers upwind and downwind of the source in question and utilizing the difference between these sampling results to determine the quantity of dust liberated by the source. An example of area sampling from a NIOSH research project is provided next to illustrate this sampling method.

In an underground limestone mine, samplers were placed in the immediate intake and return of an underground crusher to determine the amount of dust liberated during the dumping/crushing of the limestone. Figure 2-4 illustrates these sampling locations. Samplers are placed on both sides of the entry at both sampling locations to obtain more representative measurements of the airborne dust concentrations. The concentrations from both sides of the entry are then averaged. If gravimetric samplers are used for this evaluation, the samplers must operate long enough to ensure that sufficient mass is collected during sampling.

In addition, a great number of variables that can impact dust liberation are encountered in mining operations. It is often desirable to place multiple gravimetric samplers at a single area sampling location. An average dust concentration from the multiple samplers can be calculated, increasing the confidence that the measured dust levels are representative of the true dust concentration.

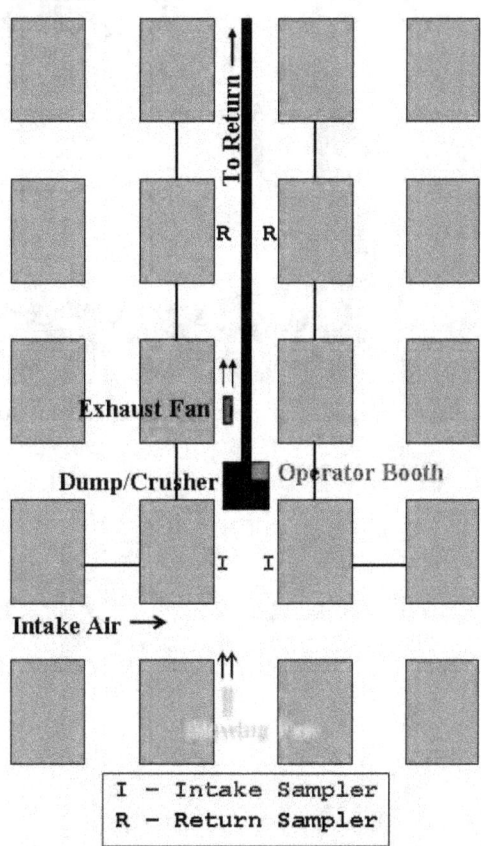

Figure 2-4. Sampling locations used to isolate dust generated at an underground crusher.

For quantifying exposures for an operator of a more mobile piece of equipment, such as a haul truck in an underground gold mine, the use of a real-time sampler such as the pDR would be beneficial to quantify dust exposure from multiple sources. Similar to Figure 2-2, the haul truck operator would be exposed to dust generated during loading of the truck, hauling to and from the dump site, and during the dumping of the load. Gravimetric and real-time samplers were mounted near the operator's compartment on the underground haul truck (see Figure 2-5) to monitor dust exposures throughout the normal haul cycle. Time study information was collected and used to separate each of these operations during analysis of the real-time data [Chekan et al. 2002]. In this manner, the average dust exposure during each segment of the haul cycle could be isolated to determine where the greatest dust exposure was realized. Figure 2-6 shows the dust concentrations from each segment of the cycle. Even though the dumping operation took the shortest time, it resulted in the highest dust levels.

Figure 2-5. Dust samplers mounted on haul truck in gold mine.

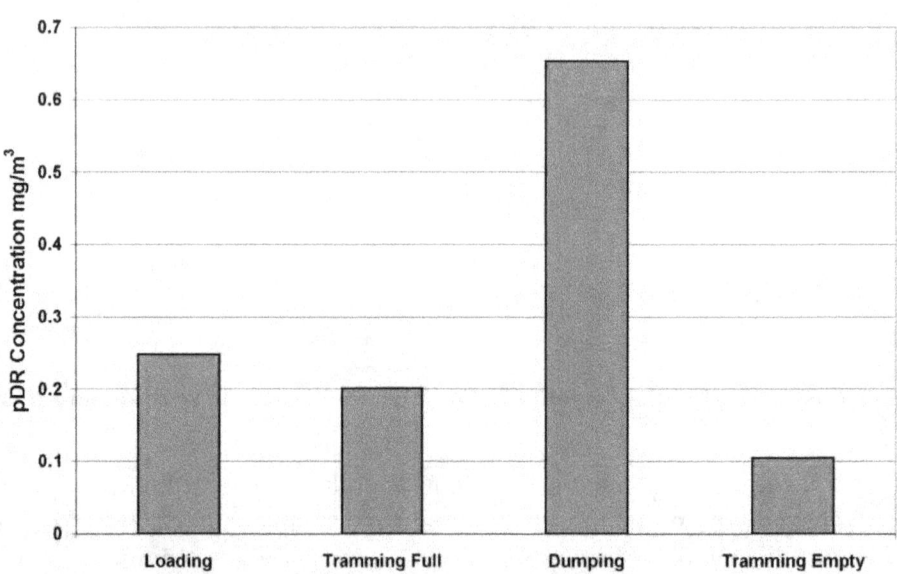

Figure 2-6. Average dust generated by each segment of sampled haul truck cycle.

If a consistent ventilation pattern is not present, it would be necessary to use multiple samplers in an effort to quantify dust from a particular source. For example, in order to quantify the amount of respirable dust that is generated by a drill at a surface mine, it would be necessary to place an array of samplers around the drill to account for dust liberated during changing wind directions.

The dust concentrations from these samplers would be averaged to quantify dust liberation around the drill. It would also be necessary to place a background dust sampler far enough away from the drill to monitor ambient dust levels. The dust levels from the ambient sample would be subtracted from the drill samples that have been averaged to determine dust liberated by the drill. Figure 2-7 shows sampling locations around a surface drill.

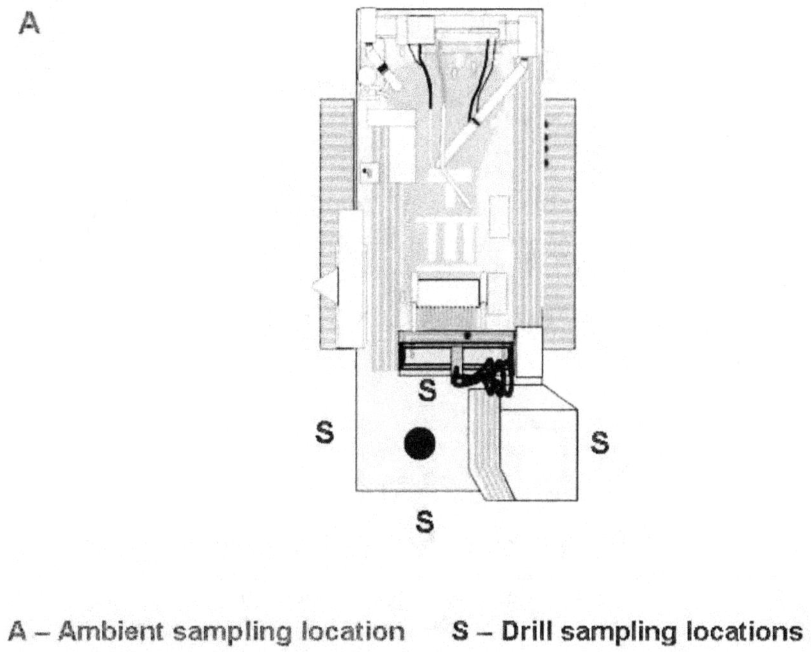

A – Ambient sampling location   S – Drill sampling locations

Figure 2-7. Sampling locations around surface drill.

After identifying the most significant dust sources, appropriate dust controls should be selected and implemented. In order to determine the impact of the additional controls, sampling would once again be conducted. Typically, an A-B comparison would be needed to quantify the impact of added control technologies. The A portion of the test would be completed under original operating conditions (e.g., mining equipment and methods, production, geologic conditions) and used to establish baseline dust levels. The control technology of interest would then be installed and the B portion of the testing completed. The most valid comparisons can be made if the operating conditions do not change between the A and B segments of testing. The difference in dust levels measured for each test condition would be calculated to quantify the effectiveness of the installed control.

REFERENCES

Chekan GJ, Colinet JF, Grau RH III [2002]. Silica dust sources in underground metal/nonmetal mines—two case studies. In: SME Transactions, Vol. 312. Littleton, CO: Society for Mining, Metalury, and Exploration, Inc., pp 187–193.

NIOSH [1994]. NIOSH manual of analytical methods. Cincinnati, OH: U.S. Department of Health and Human Services, Centers for Disease Control and Prevention, National Institute for Occupational Safety and Health, DHHS (NIOSH) Publication No. 94-113.

NIOSH [2003]. NIOSH manual of analytical methods, 4th ed., 3rd supplement. Cincinnati, OH: U.S. Department of Health and Human Services, Centers for Disease Control and Prevention, National Institute for Occupational Safety and Health, DHHS (NIOSH) Publication No. 2003-154.

NIOSH [2006]. Laboratory and field performance of a continuously measuring personal respirable dust monitor. By Volkwein JC, Vinson RP, Page SJ, McWilliams LJ, Joy GJ, Mischler SE, Tuchman DP. Cincinnati, OH: U.S. Department of Health and Human Services, Centers for Disease Control and Prevention, National Institute for Occupational Safety and Health, DHHS (NIOSH) Publication No. 2006-145.

Parobeck PS, Tomb TF [2000]. MSHA's programs to quantify the crystalline silica content of respirable mine dust samples. 2000 SME Annual Meeting, Preprint 00-159, 5 p.

Thermo Scientific [2008]. Model pDR-1000AN/1200 personal DATARAM instruction manual. Waltham, MA: Thermo Scientific, pp. 35-36.

CHAPTER 3. CONTROLLING RESPIRABLE SILICA DUST IN UNDERGROUND STONE AND METAL/NONMETAL MINES

By Gregory J. Chekan

The health hazards associated with overexposure to respirable crystalline silica dust in the mining industry have been well documented. Studies of molybdenum, lead, and gold miners in a Colorado mining community found that silica exposure was strongly associated with silicosis prevalence rates, with 13% silicotics among those with an average exposure of 0.025–0.05 mg/m^3, 34% among those with exposures > 0.05–0.1 mg/m^3, and 75% among those with exposures > 0.1mg/m^3 [Steenland and Brown 1995; Kreiss and Zhen 1996]. An analysis of the Mine Safety and Health Administration (MSHA) compliance dust sampling data has shown that a high percentage of samples with more than 1% silica from underground stone, metal, and nonmetal mines exceeded the applicable permissible exposure limit (PEL). For inspector samples collected from 2004 through 2008, over 17% of samples exceeded the PEL. High risk occupations that had samples over the PEL include crusher operator at 36%, front end loader operator at 16%, and truck driver at 11% [MSHA 2008].

The stone and metal/nonmetal mining industry encompasses many types of commodities. The potential for respirable silica dust exposure to workers in the stone and metal/nonmetal mining industry is related to the percentage of silica in the product being mined or processed. For crushed and broken stone or dimension stone, silica percentages are on the high end, with sandstones and granites averaging 70% to 90%. On the low end are limestones, averaging 20% to 30%. For all metal/nonmetal ores, silica percentages average from 5% to 20% [USBM 1992]. Therefore, airborne concentrations of silica dust are dependent upon the silica percentage in the rock and ore being mined. Each commodity has common dust sources related to the mining cycle, which includes drilling, blasting, loading, hauling, and crushing.

The purpose of this chapter is to address best practices in controlling respirable silica dust in underground stone and metal/nonmetal mines. Dust control methods commonly used in underground operations can be divided into three distinct areas: (1) the application of local and mine-wide ventilation systems to dilute, transport, and remove dust from the ambient air and direct dust away from workers, (2) the isolation of workers from airborne dust using dust filtration systems on enclosed cabs and booths, and (3) the capture of dust after generation using water sprays and wetting agents to mitigate dust entrainment.

This chapter addresses best practices for respirable silica dust control generated from four primary dust sources: (1) crushing facilities, (2) production shots, (3) mucking operations, and (4) drilling.

CRUSHING FACILITIES

Sampling surveys have shown that underground crushing facilities, which include the dump, the crushers, and the associated conveyor belts and transfer points, can be a significant source of silica dust generation. Airborne silica concentrations can be extremely high depending on the bulk content of silica in the rock and crusher production capacity. At one limestone mine, with rock composed of 30% silica by weight and the crusher operating at 1,000 tons per hour, the silica concentration measured directly above the crusher jaws was 1.8 mg/m^3 [Cheken and Colinet 2002; Chekan et al. 2003]. Occupations typically exposed to silica dust from this source obviously include the crusher operators and truck drivers, and also the mechanics, cleanup men, and laborers whose tasks require them to work in this area. Several methods for reducing worker exposure to silica dust at crusher locations are recommended, as follows:

- **Isolate the facility from the general mine air circuit.** Dust generated from this source can be adequately contained using brattice or permanent stoppings to isolate the entire facility (dump, crusher, and belt). Booster fans using blowing ventilation should be positioned in key locations to increase airflow around the facility and dilute and transport dust away from the location to a return entry. Booster fans may be either axial vane or propeller, but recent studies have shown that propeller fans dilute and transport dust more effectively, especially in large-opening mines [Chekan et al. 2006]. Figure 3-1 shows the plan view of a limestone crusher facility isolated from the other mine developments using stoppings and a blowing propeller fan to move dust-laden air into the return.

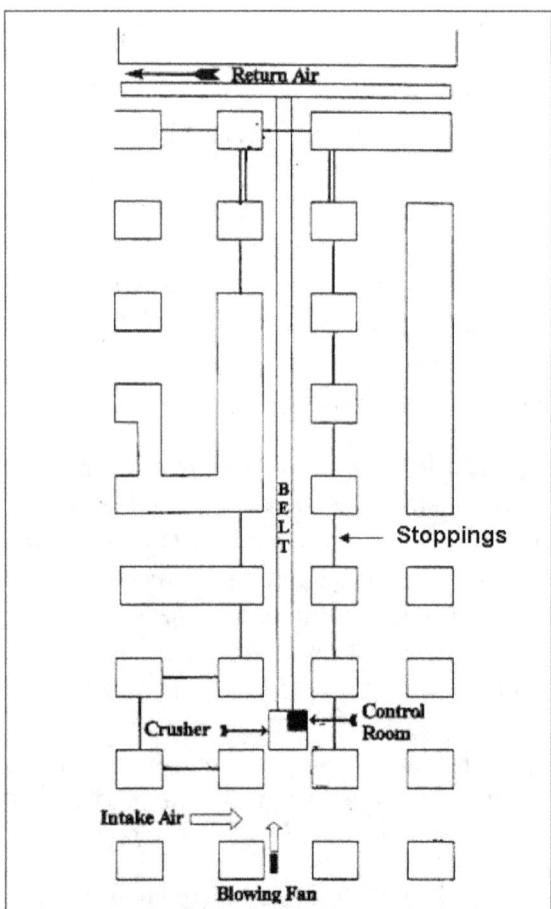

Figure 3-1. Typical method used to isolate crushing facility from mine air.

- **Ventilate with a closed ventilation system**. A closed ventilation system, where a plenum is located under the crusher, may be required in cases where the facility cannot be isolated and dust cannot be directed to the return entries. Air is exhausted from under the plenum, creating an indraft at the crusher jaws to capture the dust. The dust-laden air is then directed to a nearby return, a bag house, or a fan-powered dust collector where it is captured by filters and the clean air can be discharged into the mine air [NIOSH 2003a]. Figure 3-2 shows a conceptual approach to control crusher dust in a stone mine using a closed ventilation system.

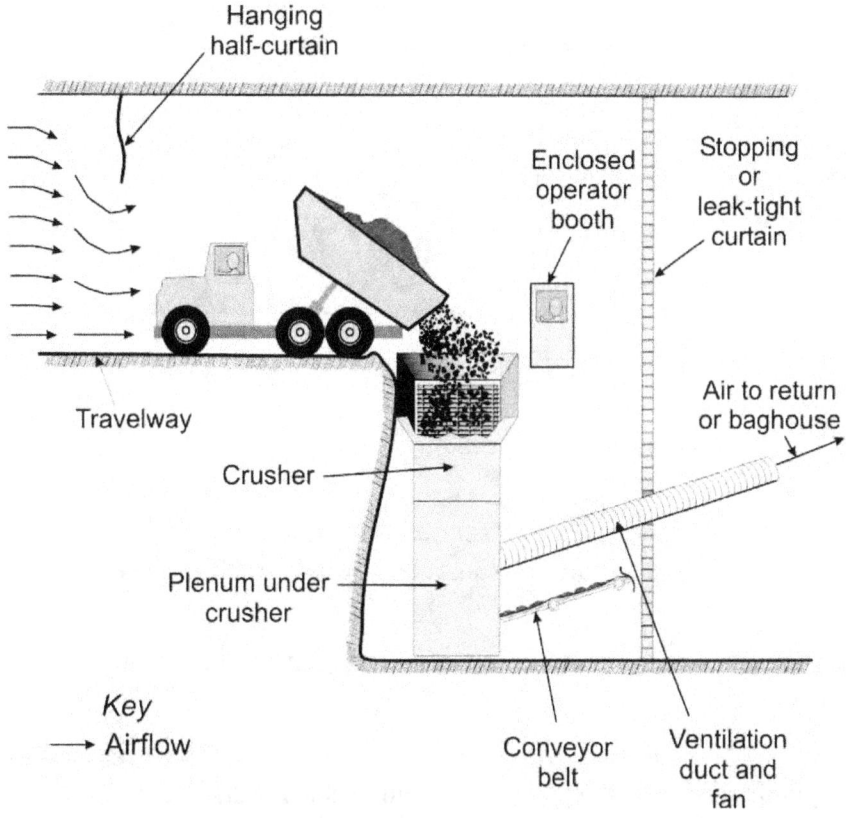

Figure 3-2. Closed ventilation system using plenum.

- **Use filtration/pressurization systems in mobile equipment cabs and operator booths**. As mining equipment ages, many of the original components on the cab enclosure deteriorate through normal operation in harsh mine environments. As a result, the effectiveness of the air filtration system and cab seals is lessened and the protection initially afforded to operators is compromised, possibly exposing them to elevated levels of respirable silica dust. NIOSH has worked with a number of manufacturers to develop cost-effective methods to improve both filtration effectiveness and cab integrity on these older cabs with the goal of reducing silica dust levels inside the cabs. Research results show dust levels inside upgraded cabs were reduced from 65% to 95% when compared to levels outside the cabs [NIOSH 2008]. Retrofit options from several manufactures are available for both enclosed cabs and booths. Figure 3-3 shows an effective design of an enclosed filtration and pressurization system.

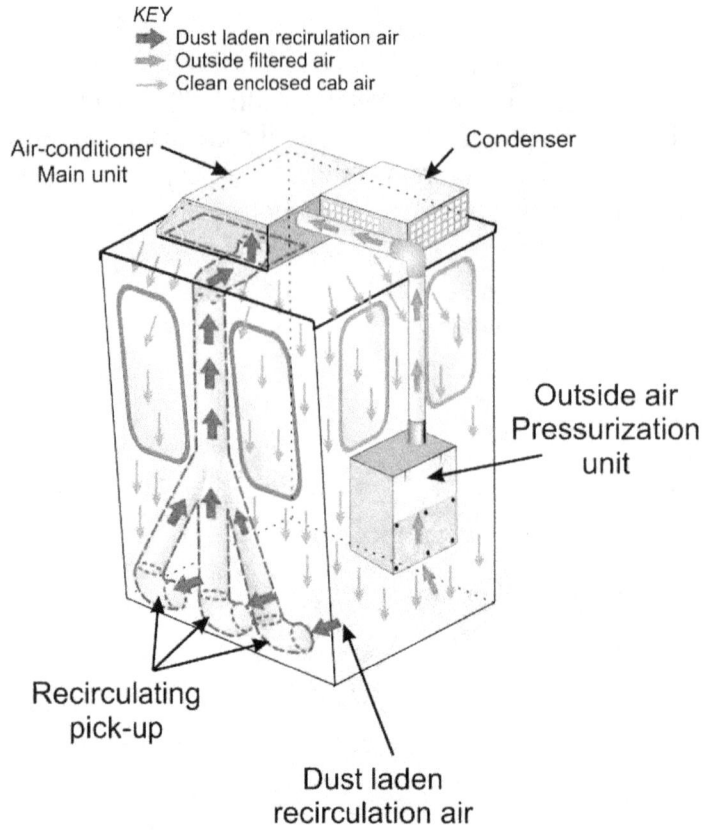

Figure 3-3. Filtration and pressurization system components and design.

- **Consider five key factors for maintaining and operating enclosed cabs and booths.**

    (1) Ensure good cab enclosure integrity to achieve positive pressurization against wind penetration into the enclosure. Studies show that significant improvements in cab protection factors were achieved when cab pressures exceeded 0.01 inches of water gauge [Cecala et al. 2005].

    (2) Utilize high-efficiency respirable dust filters on the intake air supply into the cab. Filter efficiency performance specifications used in the field were 95% or greater on respirable-sized dusts. Laboratory experiments showed an order of magnitude increase in cab protection factors when using a 99%-efficient filter versus a 38%-efficient filter on respirable-sized particles [NIOSH 2008].

    (3) Use an efficient respirable dust recirculation filter. All the cab field demonstrations used recirculation filters that were 95% efficient, or greater, on respirable-sized dusts. Laboratory experiments showed an order of magnitude increase in cab protection factors when using an 85%- to 94.9%-efficient filter on respirable-sized dusts as compared to using no recirculation filter [NIOSH 2008]. Laboratory testing also showed that the time for interior cab concentration to decrease and reach stability after the cab door is closed was cut by more than half when using the recirculation filter.

(4) Minimize dust sources in the cab by using good housekeeping practices, such as periodically cleaning soiled cab floors, using a sweeping compound on the floor, or vacuuming dust from a cloth seat [NIOSH 2001b]. Also, relocate heaters that are mounted near the floor. These units have been shown to blow air across soiled cab floors and increase dust levels inside the cab [NIOSH 2001a].

(5) Keep doors closed during equipment operation. One study showed a ninefold increase in dust concentrations inside the cab when doors were frequently opened during the sampling period [Cecala et al. 2007].

- **Use canopy air curtains**. In many underground mines, operator enclosures cannot be used due to various mining or operational parameters. An alternative option for operators in open cabs and crusher compartments is a canopy air curtain, which filters and blows clean air over the operator's breathing zone (Figure 3-4) [Goodman et al. 2006; Goodman and Organiscak 2001]. In one case study, NIOSH research has shown that the primary dust source for load-haul-dump operators without enclosed cabs occurred while dumping at the crusher. Dumping accounted for 34% of the operator's silica exposure, despite being the shortest segment of the haulage cycle [Chekan and Colinet 2002].

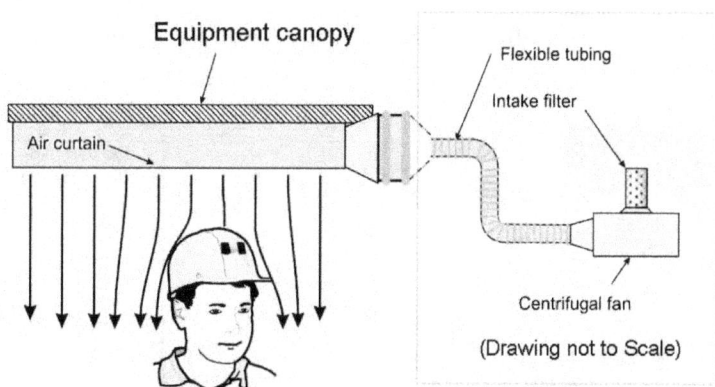

Figure 3-4. Canopy air curtain blows filtered air over worker.

PRODUCTION SHOTS

In underground stone and metal mines, production shots generate a considerable volume of dust and can be a distinct point source of respirable silica dust [Chekan et al. 2004]. Gravimetric filter samples collected 100 feet from the faces being shot showed that silica can account for over 10% of the respirable dust sample by weight, reaching concentrations as high as 0.1 mg/m^3 [Chekan and Colinet 2002]. In large mine openings, low air velocities (< 25 fpm) are common because of the large open-space volume and the extremely low airflow resistance [Krog and Grau 2006]. As a result, airflow in the entries can be stratified, or the direction of airflow can be readily affected by the movement of mine equipment. The respirable dust that becomes airborne after the production shot will remain entrained in the air and circulate with the general airflow patterns in the mine. Typically, several faces are shot at the same time, usually during an off-shift with no personnel in the mine. However, if adequate ventilation is not present or air recirculation is occurring, significant levels of silica dust can remain at the mine face and in the general mine atmosphere when workers return to begin the production cycle.

To effectively remove and reduce the retention time of silica dust after production shots, three key design parameters should be included in the mine ventilation plan:

- **A main mine fan used to establish air circuits on a mine-wide scale.** These fans include axial vane fans, jet fans, and more recently low-pressure, high-volume propeller fans. Depending on mine size, minimum air volumes of 250,000 cfm are required to adequately ventilate and maintain air velocities necessary to remove dust. Main mine fans should be mounted at the bulkhead and operated in the exhaust mode [Krog and Grau 2006; Grau et al. 2002].

- **Permanent or brattice stoppings installed in key locations throughout the mine to more efficiently direct and control the airflow.** Incorporating a stopping line into the ventilation plan using a combination of long pillars, permanent stoppings (metal/block) or temporary stoppings (brattice/curtain) has been shown to significantly improve airflow in main entry developments by providing a directional flow of air which did not exist before the systems were installed [Grau et al. 2006; Timko and Thimons 1987]. Studies showed that stopping lines reduced the retention time of dust generated by production shots and decreased the length of time for the dust to travel from the shot location to the main mine fans [Chekan et al. 2004]. Figure 3-5 shows an example of shot dust exiting the mine as recorded by a real-time dust monitor.

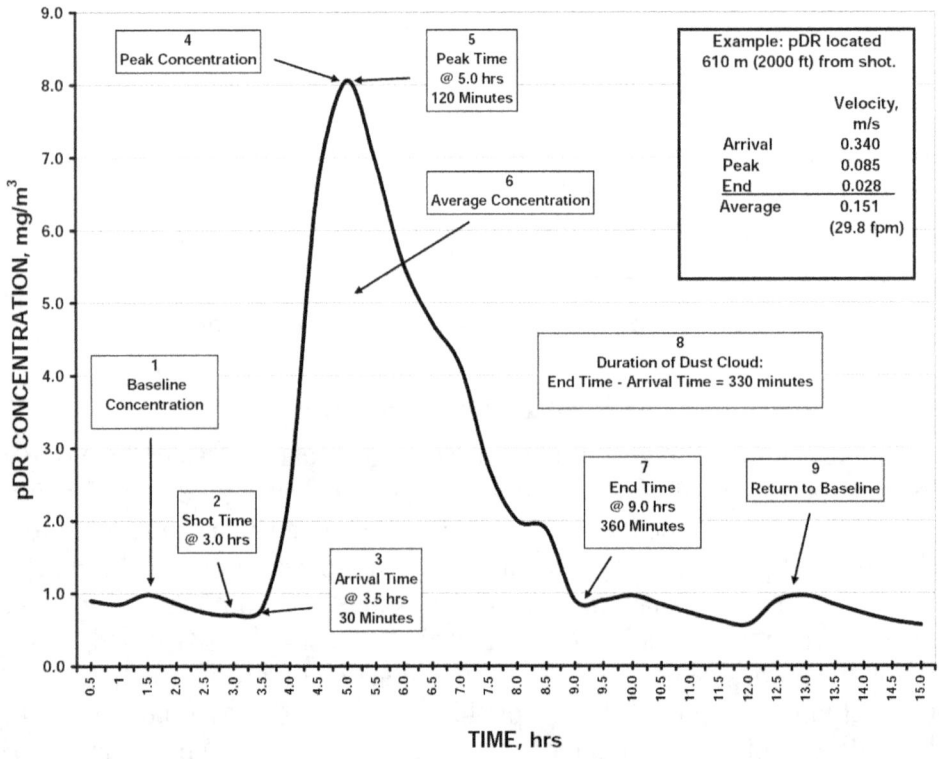

Figure 3-5. Movement of dust after production shot as recorded by real-time dust sampler.

- **Booster fans to improve local ventilation.** Booster fans, either axial vane or propeller, should be in operation near production shot locations to move dust to the primary ventilation circuit where the main mine fan can remove the dust from the mine [Chekan et al. 2004]. Permanent installation of electric axial vane fans should be located so that they blow through the fresh air stream and assist the main mine fan [Krog and Grau 2006]. Because of their mobility, diesel-powered propeller fans can be positioned closer to the shot location and oriented to produce turbulence to dilute and transport dust to the primary ventilation circuit [Chekan et al. 2006]. Figure 3-6 shows an axial vane and a diesel-powered propeller fan used to assist ventilation and remove dust near production faces.

Figure 3-6. Axial vane fan (left) versus propeller fan (right).

MUCKING OPERATIONS

Hard rock mining requires drilling and shooting of faces to produce a "muck" which is loaded and hauled using different types of production vehicles depending on commodity and mining type. Production equipment is usually diesel-powered, and vehicle cabs may be either enclosed or open depending on commodity. For instance, limestone and granite mines generally use the room-and-pillar mining method, with entry widths ranging from 30 to 60 feet and entry heights on development ranging from 20 to 45 feet. Production equipment includes large front-end loaders and 50- to 100-ton capacity trucks with enclosed cabs.

Gold and other metal operations may use sublevel caving, long-hole open stoping, or cut-and-fill methods with entries ranging from 15 to 20 feet wide and 12 to 15 feet high. Open cab load-haul-dump vehicles and muckers are commonly used in the above operations. Local ventilation and water application to the muck pile are the primary means of dust control during the loading and hauling cycle.

In mucking operations, several dust control methods should be considered to lower airborne levels of silica dust:

- **Establish an air circuit and keep fans as close to the loading area as possible.** Dead-end entries and stopes are difficult to ventilate and they create conditions where exposure to silica dust is most prevalent. In stone mines, booster fans located in key locations are commonly used to improve local ventilation and provide a more direct and controlled volume of airflow. Using a combination of booster fans in both the blowing and exhaust mode will provide both turbulent air to dilute dust and an air circuit to sweep the face and remove airborne particulate. In metal operations, blowing and exhaust systems using ventilation tubing directed into the dead-entry are more applicable. A blowing system delivering 10,000 cfm and kept within 100 feet of the face is required to provide adequate dilution. For exhaust systems to effectively transport dust-laden air into return entries, the tubing needs to be kept within 10 feet of the dust source for adequate dust capture [NIOSH 2003b]. Figure 3-7 shows a typical fan set-up for ventilating a dead-end entry in a limestone mine.

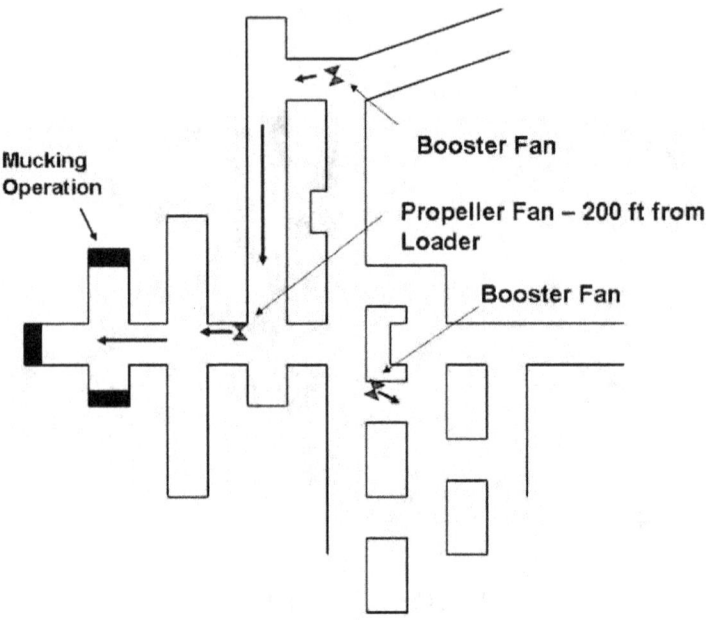

Figure 3-7. Fans positioned to ventilate dead-end entries.

- **Keep muck wet when loading.** The amount of water applied to the muck pile will differ between commodities, depending on the acceptable amount of moisture allowed during processing. Keeping the muck wet to reduce airborne dust levels while loading is a widely accepted practice, and studies have shown it also reduces silica dust levels in gold mining [Chekan 2002]. Silica generation in stopes that had a wet muck was 28% less than that produced by a dry muck. Silica exposures for load-haul-dump (LHD) machine operators were impacted during the loading and dumping activities with wet muck, with 32% and 35% dust reductions, respectively.

- **Control haul road dust.** The most common method of controlling haul road dust is surface wetting with water, but other dust control methods include adding hygroscopic salts, surfactants (commonly referred to as wetting agents), soil cements, bitumens, or films (polymers) to the road surface [Organiscak and Reed 2004]. Haul road wetting with water trucks in stone mines has been demonstrated to be very effective when continual wetting is practiced. Wetting is primarily conducted on the main tram roads; however, other trucks and mine equipment using secondary tram roads can raise the silica levels in the mine atmosphere and have the potential for exposing other mine workers conducting tasks unrelated to the production cycle.

DRILLING

Wet drilling has been shown to be effective in controlling dust and is commonly found on face drills or jumbo drills. Regular maintenance, as recommended by the drill manufacturers, should be completed to ensure proper operation and maximized protection of these systems.

Dry hole drills, where water is not used to suppress dust, are commonly used on downhole drills in preparation for shooting benches. Silica dust is generated by compressed air (bailing airflow) flushing the drill cuttings from the hole. Dry dust collection systems, incorporated into the drilling machine by the original equipment manufacturers, tend to be the most common type of dust control. Ninety percent of dust emissions with this type of system are attributed to drill deck shroud leakage, drill stem bushing leakage, and dust collector dump discharge.

- **Minimize silica dust generation and reduce levels in the ambient mine air by using the following recommended drill operating parameters:** (a) maintain a tight drill deck shroud enclosure with the ground, (b) maintain a collector-to-bailing airflow ratio of at least 3:1, (c) install a shroud on the collector dump discharge that extends close to the ground, and (d) maintain the dust collector as specified by the manufacturer [NIOSH 1998, 2005; Reed et al. 2004].

- **Use booster fans to improve local ventilation.** Booster fans should be used to improve local ventilation and remove dust from the drill site when drilling benches in areas where ventilation provided by the main mine fan is not adequately diluting and transporting dust. Studies have shown that diesel-powered propeller fans, because of their mobility and entry coverage, have favorable ventilation characteristics for this application [Chekan et al. 2006; NIOSH 2003a].

To protect drill operators from dust that escapes the controls discussed above, enclosed cabs should be used on drills and should be properly equipped with an upgraded filtration and pressurization system.

REFERENCES

Cecala AB, Organiscak JA, Zimmer JA, Heitbrink WA, Moyer ES, Schmitz M, Ahrenholtz E, Coppock CC, Andrews EH [2005]. Reducing enclosed cab drill operator's respirable dust exposure with effective filtration and pressurization techniques. J Occup Environ Hyg *2*(1):51–63.

Cecala AB, Organiscak JA, Zimmer JA, Moredock D, Hillis M [2007]. Opening door on drill cab during non-drilling can significantly increase operator's dust exposure. Rock Prod J *110*(10):29–32.

Chekan GJ, Colinet JF [2002]. Silica dust sources in underground limestone mines. In: Proceedings of the Thirty-Third Annual Institute on Mining Health, Safety and Research. Blacksburg, VA: Virginia Polytechnic Institute and State University, Department of Mining and Minerals Engineering, pp. 55–70.

Chekan GJ, Colinet JF, Grau RH III [2003]. Silica dust sources in underground metal/nonmetal mines. Trans Soc Min Met Explor *1*(312):187–193.

Chekan GJ, Colinet JF, Grau RH III [2004]. Evaluating ventilating air movement in underground limestone mines by monitoring respirable dust generated from production shots. In: Proceedings of the 10th U.S./North American Mine Ventilation Symposium, Anchorage, AK, May 16–19, pp. 221–232.

Chekan GJ, Colinet JF, Grau RH III [2006]. Impact of fan type for reducing respirable dust in an underground limestone crushing facility. In: Proceedings of the 11th North American/Ninth U.S. Ventilation Symposium, University Park, PA, June 5–7, pp. 203–210.

Goodman GVR, Organiscak JA [2001]. Laboratory evaluation of a canopy air curtain for controlling occupational exposures of roof bolters. In: Proceedings of the 7th International Mine Ventilation Congress, Krakow, Poland.

Goodman GVR, Beck TW, Pollock DE, Colinet JF [2006]. Emerging technologies control respirable dust exposures for continuous miner and roof bolter personnel. In: Proceedings of the 11th North American/Ninth U.S. Ventilation Symposium, University Park, PA, June 5–7.

Grau RH III, Mucho TP, Robertson SB, Smith AC, Garcia F [2002]. Practical techniques to improve the air quality in underground stone mines. In: Proceedings of the 9th North American/U.S. Ventilation Symposium, Kingston, Ontario, June 8–12, pp.123–129.

Grau RH III, Krog RB, Robertson SB [2006]. Maximizing the ventilation of large-opening mines. In: Proceedings of the 11th North American/Ninth U.S. Ventilation Symposium, University Park, PA, June 5–7, pp.53–59.

Kreiss K, Zhen B [1996]. Risk of silicosis in a Colorado mining community. Am J Ind Med 30:529–539.

Krog RB, Grau III RH [2006]. Correct fan selection for large opening mines: axial, vane or propeller fans—which to choose. In: Proceedings of the 11th North American/Ninth U.S. Ventilation Symposium, University Park, PA, June 5–7, pp.535–542.

MSHA [2008]. MSHA Standardized Information System, Arlington VA: U.S. Department of Labor, Mine Safety and Health Administration.

NIOSH [1998]. New shroud design controls silica dust from surface mine and construction blast hole drills. By Page SJ, Organiscak JA, Flesch JP, Hagedorn RT. Cincinnati, OH: U.S. Department of Health and Human Services, Centers for Disease Control, National Institute for Occupational Safety and Health, DHHS (NIOSH) Publication No. 98-150.

NIOSH [2001a]. Technology News 486: Floor heaters can increase operator's dust exposure in enclosed cabs. U.S. Department of Health and Human Services, Centers for Disease Control and Prevention, National Institute for Occupational Safety and Health.

NIOSH [2001b]. Technology News 487: Sweeping compound application reduces dust from soiled floors within enclosed operator cabs. U.S. Department of Health and Human Services,

Centers for Disease Control and Prevention, National Institute for Occupational Safety and Health.

NIOSH [2003a]. Dust control in stone mines. By Kissell FN, Chekan GJ. In: Handbook for dust control in mining. Cincinnati, OH: U.S. Department of Health and Human Services, Centers for Disease Control, National Institute for Occupational Safety and Health, DHHS (NIOSH) Publication No. 2003-147, pp. 57–72.

NIOSH [2003b]. Underground hard-rock dust control. By Kissell FN, Stachulak JS. In: Handbook for dust control in mining. Cincinnati, OH: U.S. Department of Health and Human Services, Centers for Disease Control, National Institute for Occupational Safety and Health, DHHS (NIOSH) Publication No. 2003-147, pp. 83–96.

NIOSH [2005]. Technology News 512: Improve drill dust collector capture through better shroud and inlet configurations. By Organiscak JA, Page SJ. U.S. Department of Health and Human Services, Centers for Disease Control and Prevention, National Institute for Occupational Safety and Health, DHHS (NIOSH) Publication No. 2006–108.

NIOSH [2008]. Key design factors of enclosed cab dust filtration systems. By Organiscak JA, Cecala AB. Cincinnati, OH: U.S. Department of Health and Human Services, Centers for Disease Control, National Institute for Occupational Safety and Health, DHHS (NIOSH) Publication No. 2009-103.

Organiscak JA, Reed WR [2004]. Characteristics of fugitive dust generated from unpaved mine haulage roads. Int J Surf Min Reclam Environ 18(4):236–252.

Reed WR, Organiscak JA, Page SJ [2004]. New approach controls dust at the collector dump point. Eng Min J July:29–31.

Steenland K, Brown D [1995]. Silicosis among gold miners: exposure-response analyses and risk assessment. Am J Pub Health 85(10):1372–1377.

Timko RJ, Thimons ED [1987]. Damage resistant brattice stoppings in mine with large entries. Eng Min J 188(5):34–36.

USBM [1992]. Crystalline silica primer. Washington, DC: U.S. Department of the Interior, U.S. Bureau of Mines.

CHAPTER 4. CONTROLLING RESPIRABLE SILICA DUST IN MINERAL PROCESSING OPERATIONS

By Andrew B. Cecala

The U.S. surface metal/nonmetal mining industry is composed of a wide range of different mineral types and processes. In 2007, there were 12,270 mines (256 underground, 12,014 surface) in the metal, nonmetal, stone, and sand and gravel industries, employing 149,177 miners (17,862 underground, 131,315 surface) [MSHA 2009]. A significant portion of these miners were working in mineral processing operations as the ore removed from the pit or quarry was processed. Throughout the mineral processing cycle, mined ore goes through a number of crushing, grinding, cleaning, drying, and product-sizing sequences as it is processed into a marketable commodity. Because these operations are highly mechanized, they are able to process high tonnages of ore. This in turn can generate large quantities of dust, often containing elevated levels of respirable crystalline silica, which can be liberated into the work environment.

The Mine Safety and Health Administration's (MSHA) respirable quartz personal dust samples from 2003–2007 verify that there were a significant number of over-exposures in many job classifications within mineral processing operations. Table 4-1 shows the job classification and the percentage of samples taken that exceeded the permissible exposure limit (PEL) over this time period.

Table 4-1. Percent of samples exceeding PEL for select occupations.

Job Classification	Samples exceeding PEL (percent)
Underground Crusher Operator	36
Stone Polisher/Cutter	33
Bagging Operator	27
Belt Cleaner	25
Hammer Mill Operator	23
Kiln/Dryer Operator	19
Cleanup Man	19
Underground Frontend Loader Operator	18
Laborer, Bullgang	16
Drill Operator	11

The purpose of this chapter is to summarize current state-of-the-art control technology available for lowering respirable crystalline silica dust at mineral processing operations. This chapter addresses dust control techniques from the time the ore reaches the primary crusher until it is packaged into some type of shipping container to be delivered to the customer. By applying the current state-of-the-art engineering controls, methods, and techniques to lower respirable dust levels, managers, engineers, and health specialists can lower respirable dust exposures to miners and work toward the ultimate goal of eliminating silicosis and other chronic respiratory diseases from miners in this industry.

The following is a listing of sections presented in this chapter:

- Primary Dumping
- Crushing and Grinding
- Transfer Points
- Conveying
- Wet Suppression (Water Sprays)
- Local Exhaust Ventilation (LEV) Systems
- Low Velocity Transport System
- Total Mill Ventilation System
- Operator Booths, Control Rooms, Enclosed Cabs
- Screening
- Packaging/Bagging Product for Shipment
- Clothes Cleaning System
- Background Issues

PRIMARY DUMPING

Ore is normally loaded into haul trucks from the pit or quarry and driven to the primary crusher location. This ore is then either dumped directly into the primary hopper, which feeds the primary crusher, or it is dumped in a stockpile. If it is stockpiled, a front-end loader then takes the ore product and dumps it into the primary hopper. In either case, a dust cloud is created during this dumping process. There are two dust sources that must be addressed during this primary dumping process—billowing and rollback.

Billowing. The first dust source is from dust that billows out from the hopper as the large volume of product is dumped from the truck or front-end loader in a very short time period. During the dumping process, ore is grinding on ore and creating dust. In addition, there is already a substantial amount of dust contained within the ore from blasting and haulage to the primary dump. As the air in the hopper is quickly displaced from the incoming ore during dumping, it entrains (carries along in the air current) these dust particles and billows out from the hopper.

Rollback. The second dust source is from rollback under the dumping mechanism. This rollback occurs either under the bed of the haul truck or the bucket of the front-end loader.

For a dust control system to be effective at the primary dump location, the dust generated from both billowing and rollback must be controlled. There are three methods to control the billowing of dust from the hopper (suppress, enclose, and filter) and one method to control rollback (a tire-stop water spray system).

Controlling Dust Billowing From Enclosure

- **Suppress.** Normally, the first dust control technique attempted for primary dump locations is water spray application. The general rule of thumb is to add enough moisture to the product where the weight of water added is equivalent to 1% of the processed ore [Quilliam 1974]. From this point, the percentage can be adjusted based upon the improvement gained from additional moisture versus any consequences from adding too much water. The amount of moisture that can be added at primary dump locations is normally not as sensitive as in later stages of the mineral processing cycle, thus higher rates or percentages can usually be tried. One important feature with a primary dump application is to only activate the water sprays during the actual dump cycle through the use of a photo cell or a mechanical switching device. Since the actual dump cycle is a very small portion of the total time, it is not appropriate to continually use water sprays during the idle time because this can cause clogging problems, as well as, wasting water. A delay timer should also be used in this application so that the sprays continue to operate and suppress dust for a short time period after the dump vehicle has moved away. For specifics about water spray types, refer to the "Wet Suppression" section of this chapter.

- **Enclose.** Enclosures for primary dump application normally require a custom design and are usually dependant on the type and size of dump vehicles being used. In some cases, walls can be constructed around the primary dump location to form an enclosure. The walls can be either stationary or removable, based in particular on whether maintenance work is necessary. In some cases for a removable enclosure, a breathable tarp fabric material, similar to the material used on over-the-road haul trucks, can be laid over to seal the top of the enclosure.

  Another technique gaining in popularity is the use of staging curtains (Figure 4-1). Staging curtains, also called stilling curtains, are curtains of varying lengths which physically prevent the natural tendency for dust to billow out of the primary dump as a large volume of product is dumped [Weakly 2000]. By minimizing the billowing airflow effect, the amount of dust released from the primary hopper is reduced.

  If staging curtains are not used, another option is to enclose the front of the enclosure using panels of flexible plastic stripping. This plastic stripping employs an overlapping sequence which provides for a very effective seal (Figure 4-2). One noteworthy benefit to the plastic stripping is that the panels are not damaged if contacted by the bucket of the front-end loader or the bed of the haul truck during dumping.

- **Filter.** When using an enclosure, it is also possible to incorporate a local exhaust ventilation (LEV) system to filter the dust-laden air from the hopper area. This would be most applicable when the primary dump is at a location where the dust could enter an adjoining structure or impact outside miners. The enclosure helps to contain the dust cloud that billows up from the hopper during the dumping process, but the dust remains airborne unless it is suppressed or removed. An LEV can be an effective technique incorporated to remove and filter this dust if it is properly designed and sized to the hopper. Since hoppers are usually large, a significant amount of airflow is typically required to create a negative pressure necessary to contain the dust cloud [MSA 1978]. Because of the large volume that must be ventilated in this application, using an LEV system would be a much more expensive control technique than the wet suppression technique.

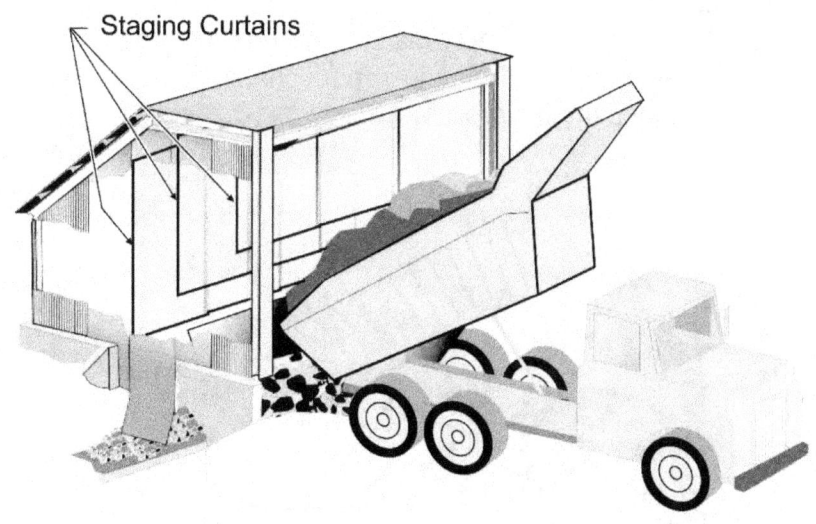

Figure 4-1. Staging curtains reduce dust billowing during dumping.

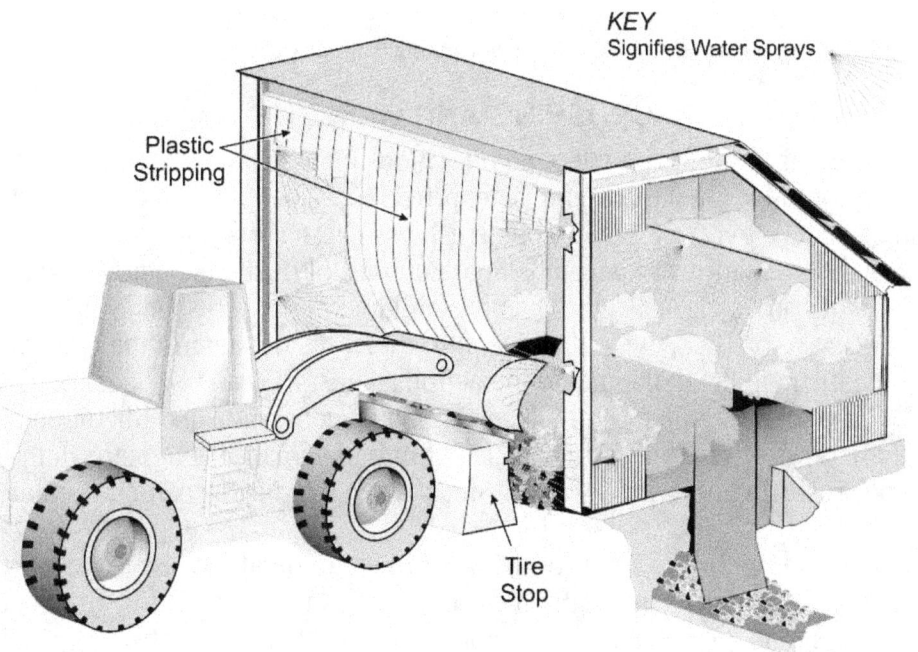

Figure 4-2. Plastic stripping holds dust inside enclosure allowing water sprays to knock down dust.

Controlling Rollback Dust

A tire-stop water spray system is recommended for reducing the dust source that causes liberated dust to rollback under the dumping mechanism. A tire stop or Jersey barrier should be positioned at the most forward point of dumping for the primary hopper. To the back side of this tire stop, a water spray manifold should be attached to knock down and force the dust, which would otherwise roll back under the dumping mechanism, to remain in the hopper. Additionally, a shield should be placed over this water spray manifold to protect it from damage from falling ore (Figure 4-3). Finally, a system should also be incorporated that allows the water sprays to only be activated during the actual dumping process.

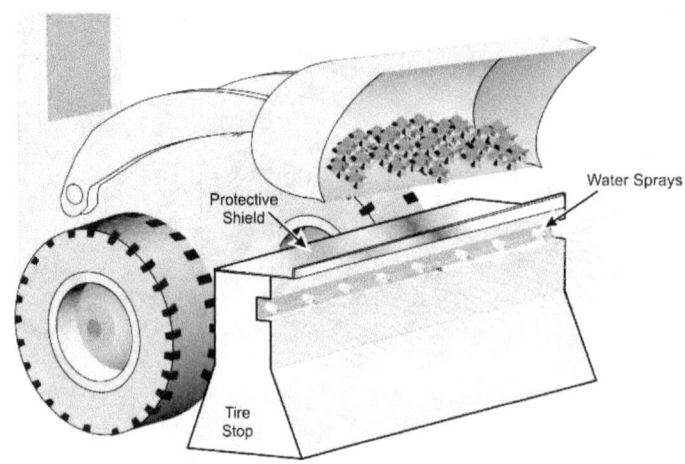

Figure 4-3. Tire-stop water spray system reduces rollback under dumping mechanism.

CRUSHING AND GRINDING

Crushing and grinding at mineral processing operations include a wide range of different types of equipment and processes. On the crushing side, primary crushers are typically jaw crushers, but may sometimes include gyratory and/or cone crushers. These crushers use compressive forces to break the ore and do not normally generate large volumes of dust. Secondary crushers may include the gyratory and cone, as well as hammermill and impact crushers. Hammermill and impact crushers use a rotating device (hammers) to thrust the ore against the outer walls of the crusher with the intent to break the ore by impaction against the outer surface. Because the ore is impacted at high velocities to induce breakage, high dust generation and liberation rates can occur from these types of crushers. After ore is fed into the crusher, it remains in the unit and continues to be crushed until it reaches a size small enough to be discharged from the unit.

Grinding and pulverizing the ore is performed later in the mineral process to reduce the product down to the smaller size ranges, normally measured in mesh sizes. Grinding mills are used to perform this process and are cylindrical, horizontal drums that rotate and have rods, balls, or pebbles inside to grind the ore down to the desired size ranges.

The two primary dust emission points of all crushing and grinding units are at the feed and discharge points. Controlling this dust by properly designing chutes or transfer points with rubber seals between stationary and moving components, as well as enclosing this area, is critical to an effective dust control plan.

Dust control for the crushing and grinding processes is normally achieved by either wet suppression or LEV systems, or a combination of both. Spraying the ore with water sprays to coat the outer surface helps to prevent dust from becoming liberated. Applying the water to the ore before it enters the crushing or grinding unit is most effective. In addition, it has been shown that the water pressure at early stages of crushing should be kept below 60 psi to avoid pressurizing and forcing dust from the feed chute enclosures [NIOSH 2003]. The amount of moisture is not as critical during the early stages of the process but should be closely evaluated as the ore enters the later stages when the finer product sizing is taking place. In these cases, full and hollow cone sprays would normally be used to wet the ore and minimize dust liberation. (See the "Wet Suppression" section for more information on water spray systems.)

When using an LEV system to capture and remove the dust from the crushing and grinding processes, a critical component to maintaining an effective system is determining the amount of air volume required to keep the process under negative pressure. As ore is fed into the crusher or grinder, it entrains air along with the product, creating a significant volume of air which must be exhausted to overcome the induction effect [MSA 1978; Yourt 1990]. The volume of exhaust air is also dependant on the effectiveness of sealing the crusher's or grinder's intake opening. By minimizing the area of the opening using belting and plastic stripping, the volume of exhaust air can be lowered while still maintaining an acceptable negative pressure necessary to contain the dust liberated during this transfer process.

One final component that must be considered in all crushing and grinding processes is maintaining a proper seal on the device. If product is observed on the floor below the device or if visible dust is seen liberating from a unit, this indicates that a hole has been created or a seal has worn out, and maintenance needs to be performed to repair the problem.

TRANSFER POINTS

Transfer points are used to move ore from one process or one piece of equipment to another. Although this seems like a simple process, significant dust generation and liberation can result from transfer points if they are not properly designed and installed. The following are some important design considerations for an effective transfer point or chute:

- Transfer chutes should be sized to allow ore to flow without clogging or jamming. A general rule of thumb is that the chute depth should be at least three times the maximum lump size to avoid clogging [Martin Marietta Corp 1987].

- The dump point of the ore should be designed to impact on a sloping bottom or a rockbox. Rockboxes are designed to allow ore product to build up so that ore contacts ore during transfer, which reduces wear and abrasion of the chute.

- Any abrupt changes in product direction or flow should be avoided.

- Fall height of ore should be minimized whenever possible through the use of rock ladders, telescopic chutes, spiral chutes, and bin-lowering chutes.

- A head enclosure should be used when transferring ore onto a conveyor. The head enclosure should be designed with strip curtains to minimize air induction into the enclosure and skirt boards to position the ore on the center of the belt.

- An LEV system should be used at transfer enclosures and chutes to capture and filter the dust from the air. These enclosures should be designed to have approximately a 250-fpm intake velocity at any opening to eliminate dust leakage from the area. To accomplish this, plastic stripping and other types of sealing systems should be used to minimize openings and maximize intake velocity. One study also recommended that the exhaust port to the LEV system be located at least 6 feet away from the transfer dump point to minimize the possibility of entraining large particles [MAC 1980].
- The exit velocity from the enclosure or chute should be kept below 500 fpm to minimize the entrainment of large particles of ore [Yourt 1990].

CONVEYING

At mineral processing operations, conveyors are the major component used to transfer ore from one process to another. A conveyor can generate significant quantities of respirable dust and be one of the greatest sources of dust emissions within an operation. There are four main areas of dust generation and liberation from conveyors:

- When ore is dumped onto the belt.
- As ore travels on the belt.
- From the underside return idlers due to carryback on the belt.
- When ore is dumped or transferred to another belt or process.

One of the challenges with conveyors is the number of belts used and the total distance traveled throughout a mineral processing plant. Some belts are located outside where dust liberation is not as critical as when they are within a facility. Another challenge particular to conveyors is their ability to generate or liberate dust while operating, whether they are loaded with ore or empty.

Controlling dust from conveyors requires constant vigilance by the maintenance staff to repair and replace worn and broken parts. There are a number of techniques to reduce dust liberation from conveyors, as follows:

- **Suppress.** When properly designed and installed, water sprays are a cost-effective method of controlling dust from conveyors. The most common and effective practice for conveyor sprays is to wet the entire width of product on the belt. The amount of moisture applied should be varied and tested at each operation to determine the optimum quantity, but 1% moisture added to product ratio is a good starting point. A number of studies have indicated that wetting the return side of the conveyor belt also helps reduce dust liberation. This is effective because it reduces dust generation from the idlers as well as at the belt drives and pulleys. In many cases, water sprays located on the top (wetting the product) and the bottom (reducing dust from the idlers) at the same application point can be an effective strategy [Courtney 1983; Ford 1973]. These locations are also beneficial from an installation and cost standpoint.

    When considering nozzle type for these suppression systems, fan spray nozzles are normally the most common design because they minimize the volume of water added for the amount of coverage. For these types of applications, it is more advantageous to locate the water sprays at the beginning of the dust source (i.e., the dump or transfer location), because as the water and ore continually mix together, the amount of wetted surface area of the ore increases, thus increasing the suppression potential and reduction in dust liberation. Using sprays at higher flow rates can sometimes create air turbulence, which makes it more difficult to contain/suppress the dust. Because of this, it is normally recommended to use more spray nozzles at lower flow rates and position them at locations closer to the ore [NIOSH 2003].

- **Enclose.** Enclosures are an effective dust control technique for many applications within mineral processing plants if they are correctly designed and installed, and this principle also applies to conveyors. Enclosures for conveyor and transfer points can be either full or partial type [Zimmer 2003]. One of the most common partial types of enclosures is with the use of skirting, which keeps the material on the belt, especially immediately after it exits a loading chute. An inclined skirting design, in which the skirting belt is angled at approximately 30 degrees from vertical, is more advantageous over a standard vertical design because of wear issues. This skirting design improves the loading of ore onto the conveyor and reduces the amount of dust generated.

    Enclosures at head and tail ends of the conveyor are a common practice because they are effective at controlling dust at these locations. Designing the proper size enclosure is a critical factor because, as the ore is dumped onto the conveyor, it entrains a measurable amount of air (venturi effect) and this can pressurize the enclosure if it is undersized. Dust curtains are another form of enclosure used to contain dust within a conveyor and are very cost effective to install. These curtains are normally installed at the head and/or tail ends of the conveyor.

    In many instances, a LEV is tied into the enclosure at these conveyor dump or transfer locations to capture generated dust. It has been shown that the takeoff port to the LEV system should be a least 6 feet from the dump point to minimize the pickup of oversized particles [MAC 1980]. The air velocity at this exhaust port should also be kept below 500 fpm to avoid the pickup of larger particles [Yourt 1990].

- **Belt scraper.** An effective method to reduce dust being liberated from conveyors is with belt scrapers. Although belt scrapers come in many different styles, types, and trade names sold by numerous commercial manufacturers, their function remains the same, which is to reduce the amount of carryback on the belt once the ore is discharged. Carryback is the material that sticks or clings to the conveyor belt after the material is discharged at the head pulley. As this material dries and passes over the return idlers, it falls from the belt and the respirable portion of this dust becomes airborne. The goal is to remove this carryback product before it is released into the air and becomes a source of contamination to the workers. When dust levels are high, a common practice is to use two or three belt scrapers at different locations in an effort to further reduce the amount of carryback material on the belt [Roberts et al. 1987].

- **Belt wash.** Some studies have shown that the oversized material is more easily removed through scraping but the smaller respirable-sized particles tend to remain adhered to the conveyor. When this occurs, a belt wash should be installed. A belt wash sprays the conveyor belt with water while simultaneously scraping it to remove the product. In a number of published studies in this area, this technique has been shown to increase the cleaning effectiveness by approximately 14% [Planner 1990].

- **Effective belt loading.** Providing an effective belt loading area helps to reduce the amount of dust generated during belt loading. The first design goal of an effective system is to reduce belt sag and vibration during ore loading. To do this, some operations are using slider-bar cradles which stabilize this area [Stahura and Marti 1995]. These cradles can be made to provide a shock-absorbing action to cushion the impact during belt loading. This goal can also be achieved with low-friction bars that provide a flat support area for loading, which minimizes belt sag. Effective belt loading should also include side shields which contain the fine-sized particles and help to minimize the liberation of dust.

Figure 4-4 shows a combination of the various techniques discussed that help enclose and minimize dust generated during conveying. This includes the use of plastic stripping to minimize the opening size, a rockbox, belt skirting, and the dust collection pick-up point. In addition, wet suppression is also shown, which will be discussed in the next section of this report.

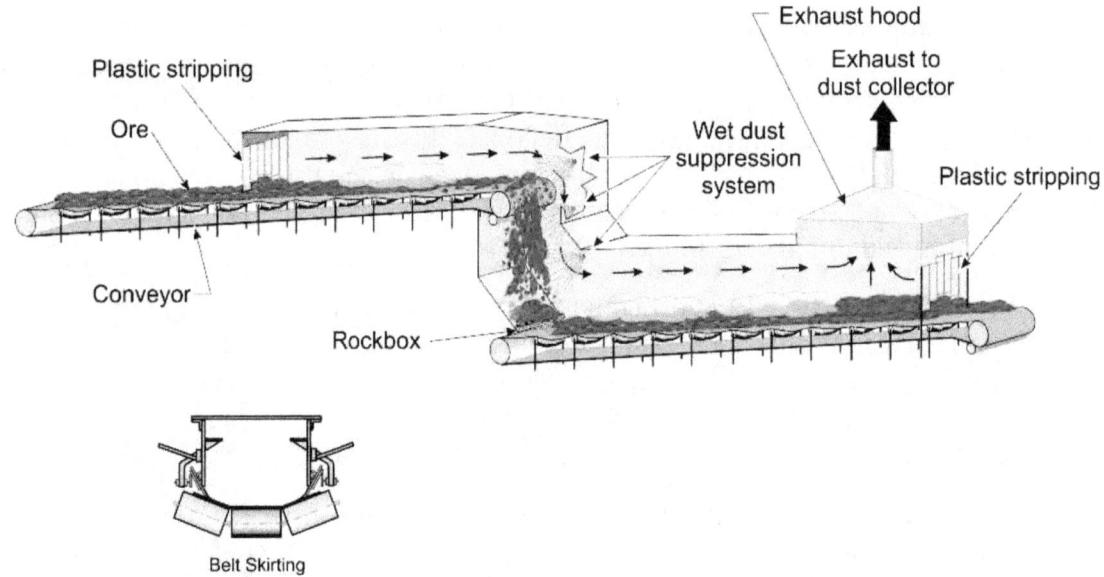

Figure 4-4. Techniques for reducing respirable dust liberation from conveyor belts and transfers.

WET SUPPRESSION

Wet suppression systems are probably the oldest and most often used method of dust control at mineral processing operations. In the vast majority of cases for mineral processing operations, the wet suppression system used is a water spray system. Although the use of water sprays is a simple technique, there are a number of factors that should be evaluated to determine the most effective design for a particular application. There are two methods to control dust using water sprays at mineral processing operations:

- Preventing dust from becoming liberated and airborne by directly spraying the ore.
- Knocking airborne dust down by spraying the dust cloud and causing the particles to collide with water droplets and fall out of the air.

Most operations use a combination of both methods in the overall dust control plan. When considering the use of a wet suppression system, some general considerations and guidelines apply:

- The effectiveness of water spray application is dependent on nozzle type, droplet size, spray pressure, spray pattern, spray angle, spray volume, spray droplet velocity, and spray droplet distribution.
- Each ore type and application point is a unique situation and needs to be evaluated separately to achieve the optimal design.
- Water evaporates and needs to be reapplied at various points throughout the process to remain effective.
- Water freezes and its use is limited during certain times of the year and in certain climates.
- Wet suppression cannot be used with all ores, especially those that have higher concentrations of clay or shale. These minerals tend to cause screens to bind and chutes to clog, even at low moisture percentages.
- Over application in the volume of moisture is a problem in all operations and can impact the equipment as well as the total process. In most cases, a well-designed suppression system will not exceed 0.5% moisture application, which is roughly equivalent to one gallon per ton of ore.
- The suppression system should be automated so that sprays are only used during times of production when ore is actually being processed. For dust knockdown, a delay timer may be incorporated into some applications to allow the suppression system to operate for a short time period after a dust-producing event.

When considering sprays, one of the primary aspects is the droplet size. When wetting the ore to keep dust from becoming airborne, droplet sizes above 100 microns should be used. In contrast, when the goal is to knock down existing dust in the air, the water droplets should be in size ranges similar to the dust particles. The intent is to have the droplets collide and attach themselves to the dust particles, causing them to fall from the air. In these cases, droplets in the range of 10 to 50 microns have been shown to be most effective [Bartell and Jett 2005]. Uniformity of wetting is also a very important issue for an effective system. By far the best dust reductions can be achieved by spraying the ore with water and then mechanically mixing the ore and water together to achieve a uniformity of wetting.

The following is a list of the spray nozzles used in the mineral processing industries and some of their defining characteristics (Figure 4-5):

- **Full-cone.** Full-cone nozzles employ a solid cone-shaped spray pattern with a round impact area that provides high velocity over a distance. They produce medium to large droplets sizes over a wide range of pressures and flows. They are normally used when the sprays need to be located further away from the dust source.

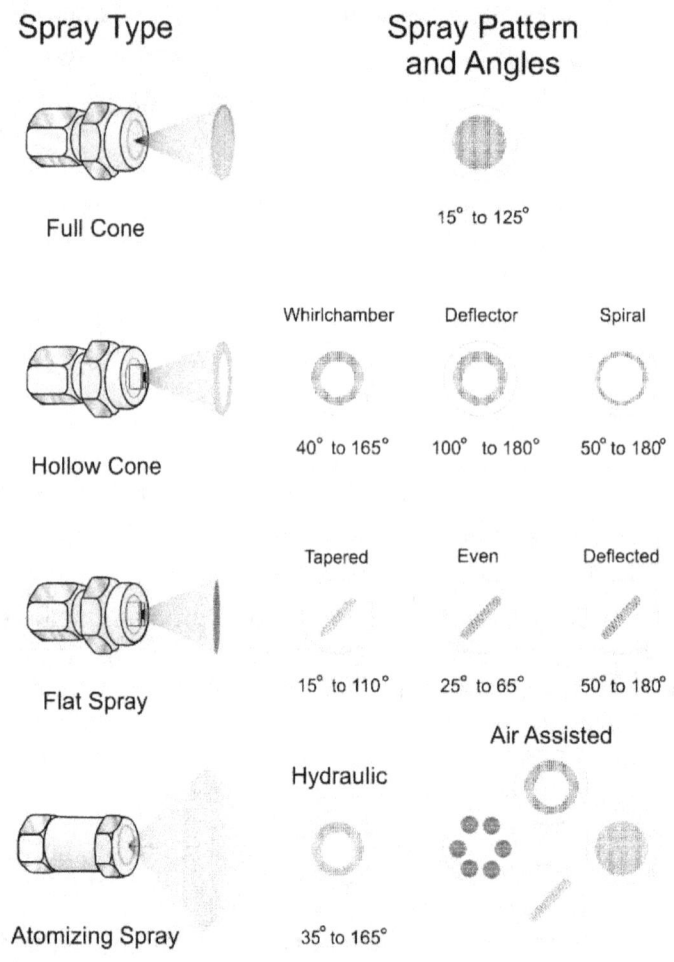

Figure 4-5. Spray nozzles commonly used in mineral processing operations.

- **Hollow-cone.** These nozzles use a circular outer ring spray pattern (hollow cone) in three different designs: whirlchamber, deflector, and spiral sprays. They produce small to medium droplet sizes.
- **Flat-fan.** The flat-fan pattern is produced in three different designs: tapered, even, and deflected type sprays. Flat-fan nozzles produce small to medium droplet sizes over a wide range of flows and spray angles and are normally located in narrow, enclosed spaces.
- **Air atomizing.** Different spray patterns are available in two different designs: hydraulic and air-assisted. Hydraulic nozzles produce fine-mist droplet sizes and have low-volume capacities. Air-assisted nozzles produce the smallest droplets of all sprays but are the most expensive because they require compressed air. To be effective, both types of air-atomizing sprays need to be located close to the dust source.

Figure 4-6 shows the airborne capture performance of the different spray nozzles performing at different operating pressures. As shown, atomizing sprays are the most efficient for dust knockdown, followed by the hollow-cone sprays. Hollow-cone sprays are a good choice for many applications in mineral processing operations because significant coverage or wetting of the ore occurs, even at low moisture percentages. They are also very beneficial because they have large orifice sizes and are less likely to clog as compared to the other nozzles. Full-cone sprays would be most applicable in the early stages of the process where the quantity of moisture added is not as critical. Flat-fan sprays are most appropriate for spraying into a narrow rectangular space because less water is wasted by spraying against an adjacent rock or metal surface.

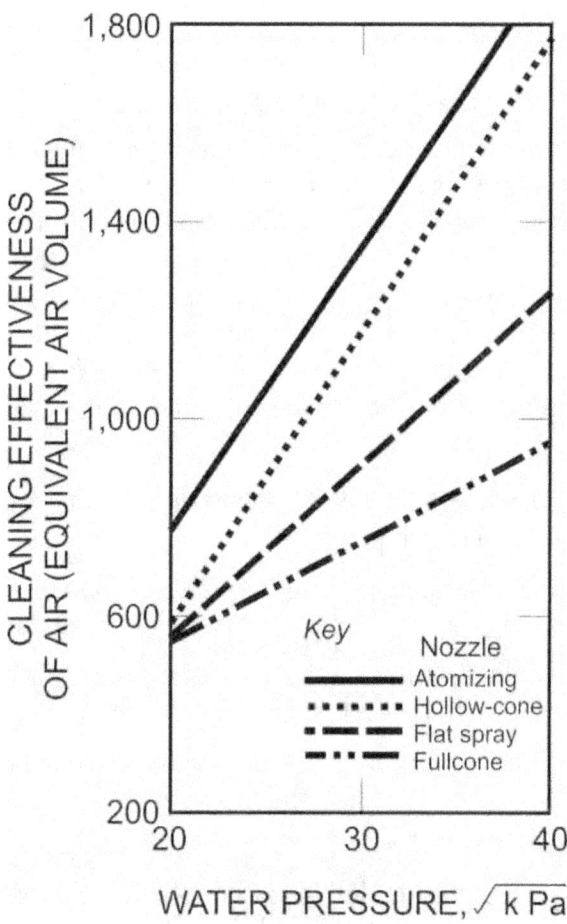

Figure 4-6. Airborne dust capture performance of four types of spray nozzles.

Surfactants, commonly called wetting agents, are sometimes used in wet suppression applications because they lower the surface tension of the water solution. This in turn increases the solution's ability to produce finer and a greater number of water droplets, while increasing the rate at which the droplets are able to wet dust particles. They use less moisture to produce the same effects as a straight water application. However, they are not very often used in the metal/nonmetal mining industry based upon the following limitations:

- Surfactants are more expensive than a typical water application.

- Surfactants may alter the properties of the mineral or material being processed.
- Surfactants can damage some equipment such as conveyor belts and seals.
- Surfactant systems require more upkeep and maintenance than typical water systems.

A fundamental consideration with any suppression system is the cleanliness of the water. If spray nozzles become plugged with sediment or debris, they render the suppression system ineffective. Since the water to be used for suppression systems at most mineral processing operations would be drawn from a settling pond, the issue of water purity is a legitimate concern. It is recommended to use some type of water filtering system to eliminate the possibility of sediment and debris clogging the water sprays. Most spraying companies offer filtering systems for this purpose. A hydrocyclone with a built-in accumulator flush should be considered in these applications [USBM 1976, 1981].

LOCAL EXHAUST VENTILATION (LEV) SYSTEMS

The most common dust control technique at mineral processing plants is local exhaust ventilation (LEV) systems. These systems capture dust at the various processes such as crushing, milling, screening, drying, bagging, and loading and then transport this dust via ductwork to a dust collection filtering device. LEV systems use a negative pressure exhaust ventilation technique in an attempt to capture the dust before it escapes from the process. By capturing the dust at the source, it is prevented from becoming liberated into the processing plant and exposing workers. This technique is most effective when a capture device (e.g., enclosure, hood, or chute) is incorporated at the dust source to maximize the collection potential. LEVs have a number of advantages:

- Ability to capture and eliminate very fine particles, which are difficult to control using wet suppression techniques.
- Providing the option of reintroducing the captured material back into the production process or discarding the material so it is not a detriment later in the process.
- Effectiveness in cold weather conditions because of not being greatly impacted by low temperatures as with the wet suppression technique.
- May be the only dust control option available for some operations whose product is hygroscopic or suffers serious consequences from even small percentages of moisture (e.g., clay or shale operations).

There are many different types of dust collection systems available for use at mineral processing operations, including electrostatic precipitators, fabric collectors, wet collectors, and dry centrifugal collections; however, an LEV system using a canister-type collector is often the best choice for the majority of the minerals processing industry. The American Conference of Governmental Industrial Hygienists (ACGIH) produces a manual entitled *Industrial Ventilation Handbook—A Manual of Recommended Practice for Design*, currently in the 26th edition. This manual provides extensive and authoritative information for designing an effective LEV system. It is not the intent of this chapter to duplicate this information, but to briefly describe a canister-type LEV system. For more information on various dust collector types and design considerations, please refer to the ACGIH *Industrial Ventilation Handbook* [ACGIH 2007].

The canister-type collector is the most recent generation of fabric-type collectors. Currently, all other fabric collectors use filter media made of felt-type collector bags. Instead, the canister-type collector uses a rigid cartridge that houses a pleated filter media (Figure 4-7). The following are some advantages of using this system:

- Various types of filter cartridges are available to meet a wide range of ore types and filtering needs.
- The filter canisters are self-cleaned in the collector unit when the internal pressure reaches a set value. This self-cleaning is performed using the conventional pulse-jet cleaning action.
- The pleated-type canister filters increase the surface area of the filter media and provide for a greater cycle time before cleaning, when compared to the typical bag-type filters.
- Workers are exposed to very low respirable dust concentrations when replacing filter canisters. The normal replacement procedure is to remove the new filter canister from the cardboard shipping package, then remove the used canister from the collector unit and place it in the cardboard shipping container to be discarded. The cardboard container lid is closed and taped shut to minimize the potential for any dust leakage.

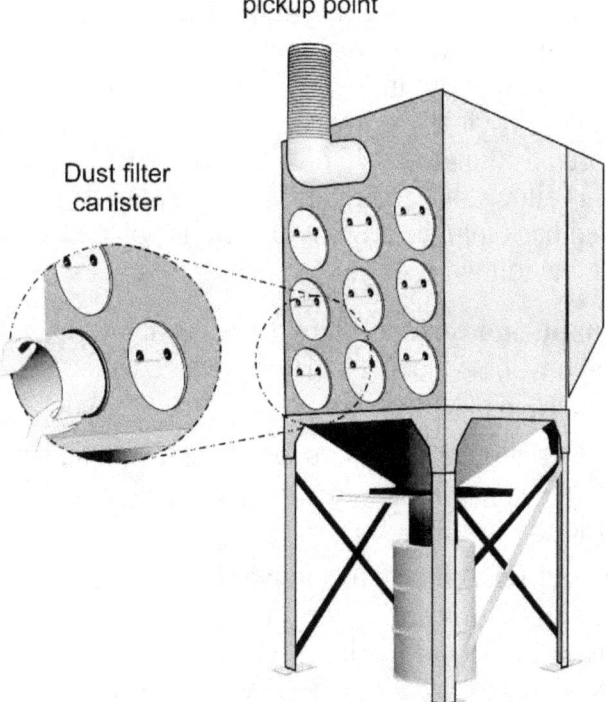

Figure 4-7. Canister-type dust collector system.

LOW-VELOCITY TRANSPORT SYSTEMS

As stated in the previous section, LEV systems are the most common dust control technique used at mineral processing plants to capture and filter dust from the air. With these systems, once the dust is captured at the source, it is carried through ductwork to the filtering unit. This section deals with a novel method of transporting dust from the collection point to the filtering unit.

For many years, the only recommended practice was to transport dust particles in a high-velocity design in which the air velocities within the duct were kept in the 3,500 to 4,000 feet per minute range. The intent with a high-velocity system is to prevent the dust particles from settling out from the airstream and clogging the duct. However, since the air velocities are so high, duct wear is a significant problem, especially at any elbows or transitions where holes are constantly developing. As holes develop, they compromise the system's performance and cause dust to leak from the ductwork and into the work environment.

A low-velocity system has been developed [Bresee 2008] where the transport velocities within the ductwork are held below 1,800 feet per minute. This section will present the advantages to the low-velocity design, which may be advantageous over the high-velocity design for some applications. This is especially true when dealing with an ore that contains silica due to the abrasiveness of the mineral. It must be noted that low velocity does not imply low airflow; in fact, hood capture velocities and negative pressure at the capture point are identical in both low- and high-velocity transport systems.

There are significant differences between high- and low-velocity system designs. In high-velocity systems, the ductwork is mainly oriented in either the horizontal or vertical format, and at the high-transport velocities, large-size dust particles are moved through the ductwork to the dust collector unit. This is not the case in a low-velocity system in which the basic principle is to only move respirable-sized particles. The low-velocity system is based upon a "sawtooth design" in which the up-slopes are positioned at a 45 degree angle and all down-slopes are oriented at 30 degrees (Figure 4-8). This allows the larger particles to fall out from the airstream, slide down the slope, and be recycled back into the process at various points. Advantages to the low-velocity transport system include the following:

- **Wear and maintenance.** Since the system is only moving particles in the respirable range at relatively low velocities, there is reduced abrasion to the duct, which allows for shorter radius and elbows to be used.
- **Energy costs.** Friction and pressure losses in this system are significantly less than in the conventional high-velocity system, making overall power requirements significantly less.
- **Availability and reliability.** The low-velocity system maintains a more consistent balance, since changes in airflow or pressure drops in ducts do not have significant impacts in the overall system. Since the system stays in balance, it does not lose its transport effectiveness as with the high-velocity transport system.
- **Product recovery.** Product losses are significantly less since larger particles are captured by the LEV system and ultimately recycled back into the process at various points in the system.

Although low-velocity transport systems have higher initial costs, the break-even point usually occurs sometime around the three-year time frame. The low-velocity transport system has been tested and utilized in operating plants, with a projected life between 15 and 25 years. This performance time frame accounts for significant cost savings over the life of the system.

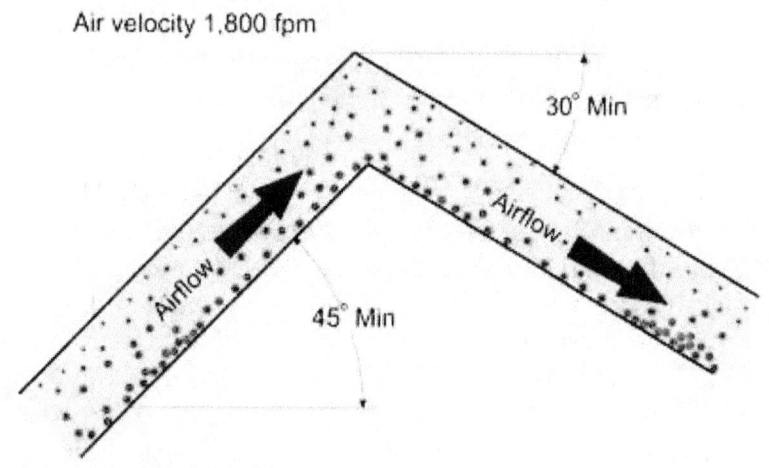

Figure 4-8. Sawtooth design for low velocity transport system.

TOTAL MILL VENTILATION SYSTEMS

Although LEV is the most common dust control technique used at mineral processing operations to capture and filter dust from the major dust sources, it is not possible to capture and control all the minor dust sources within a mineral processing operation using this same technique. As these minor sources continually generate and liberate dust over a work shift, they can have a cumulative effect and cause respirable dust concentrations to gradually increase to unacceptable levels.

The best way to address and deal with these minor dust sources is to install a total mill ventilation system (TMVS). A TMVS provides a general purging of the plant air to minimize dust throughout the entire mill building, thus lowering respirable dust levels for all workers within the structure [Cecala and Mucha 1991; Cecala et al. 1995]. The TMVS lowers respirable dust levels by using clean outside air to sweep up through a building to clear and remove the dust-laden air. This upward airflow is achieved by placing exhaust fans at, or near, the top of the structure. The size and number of exhaust fans is based upon the initial respirable dust concentration and the total volume of the structure. All the processing equipment within a mill generates heat and produces a thermodynamic chimney effect that works in conjunction with the TMVS. To be effective, a TMVS must meet three design criteria:

- **Supply of clean make-up air.** The TMVS must supply clean make-up air at the plant's base. Any outside dust sources, such as bulk loading near an air inlet location, can cause outside dust-laden air to be drawn into the plant and make the problem worse. It is critical to ensure that the make-up air is clean by controlling the air's entry location through inlets such as wall louvers, plants doors, or other openings.

- **Effective upward airflow pattern.** The TMVS should provide an effective upward airflow pattern that ventilates the entire plant and also sweeps through dust sources, work areas, and dust-laden areas. To create the most effective airflow pattern for purging the entire mill, the exhaust fans must be properly located on the roof or high outer walls, and the make-up air inlets must be located at the base of the structure.

- **Competent mill shell.** Because a TMVS uses exhaust fans to draw make-up air through the points of least resistance, the plant's outer shell needs to be intact and competent. Any unwanted openings, especially near the exhaust fans, can short-circuit the ventilation system's designed airflow pattern and reduce the effectiveness.

Figure 4-9 shows the concept of the TMVS from an outside perspective. A normal range of airflow for a TMVS would be 10–35 air changes per hour (ACPH). During the development of this technique, two different field studies were performed to document the effectiveness of a TMVS [USBM 1993]. In the first study, with a 10-ACPH ventilation system, a 40% reduction in respirable dust concentration was achieved throughout the entire mill building. In a second field evaluation, two ventilation volumes were tested—17 ACPH and 34 ACPH—and average respirable dust reductions were recorded at 47% and 74%, respectively.

The TMVS has proven to be a cost-effective system to lower respirable dust concentrations throughout an entire mineral processing structure. Not only is the initial cost of this technique inexpensive when compared to the other engineering controls, the operation and maintenance are also minimal. To further reduce their costs, operations can potentially install all the components for this system with in-house personnel. Therefore, the TMVS can be a very cost-effective system to lower respirable dust levels at mineral processing operations [Cecala 1998; Cecala and Thimons 1997; Cecala, et al. 1996].

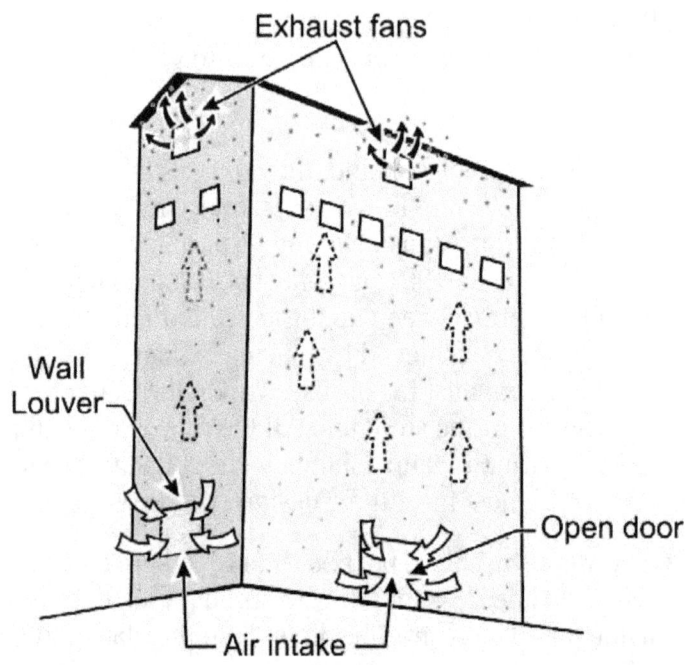

Figure 4-9. Design concept of TMVS showing clean-air intakes and dust-laden air exhausts.

OPERATOR BOOTHS, CONTROL ROOMS, ENCLOSED CABS

Frequently at mineral processing plants, workers will be located in an operator's booth, control room, or enclosed cab to give them a safe work area and to isolate them from dust sources. If these areas are properly designed, they can provide good air quality to the worker. On the other hand, if these enclosed areas are not properly designed and maintained, the air quality can deteriorate to unacceptable and unsafe levels.

The most effective technique for reducing operators' exposure to airborne dust in booths/control rooms/enclosed cabs at mineral processing operations is with filtration and pressurization systems. The most effective filtration and pressurization systems have the heating and air conditioning (HVAC) components tied in as an integral part of the system. A substantial amount of research has been performed over the past few years evaluating the air quality in enclosed cabs of surface mining equipment. This research is directly applicable to operator booths and control room dust control systems. As the research demonstrates, enclosed cabs on mobile equipment are harder to control and maintain since the moving of the equipment constantly stresses and compromises the competency of the enclosure. NIOSH recently conducted a controlled laboratory study to evaluate the key factors necessary for achieving an effective enclosure filtration and pressurization system [NIOSH 2008]. Through this laboratory and numerous field studies, the following items were identified as key components to an effective system:

- **Ensure booth/control room/cab integrity.** Effective protection factors were realized in various field studies when positive pressures between 0.01 and 0.40 inches of water gauge were achieved within the booth/cab because of good enclosure integrity. These pressures correspond to wind velocity equivalents of 4.5 to 29 miles per hour and prevent against wind forcing dust laden air into the enclosure.

- **Use high-efficiency filters on intake air.** Only intake filters with an efficiency of 95% or greater were used during field studies [Cecala et al. 2004, 2005a; Chekan and Colinet 2003; Organiscak et al. 2004]. Laboratory experiments showed an order of magnitude increase in protection factors when using a 99%-efficient filter versus a 38%-efficient filter on respirable-sized particles [NIOSH 2007].

- **Use an efficient recirculation filter.** All the field evaluations used recirculation filters that were 95% or greater on respirable-sized dusts [Cecala et al. 2004, 2005a; Chekan and Colinet 2003; Organiscak et al. 2004]. Laboratory experiments showed a tenfold increase in protection factors when using an 85%- to 94.9%-efficient filter on respirable-sized dusts as compared to using no recirculation filter [NIOSH 2007]. Laboratory testing also showed that the time needed for the interior to stabilize after the door was closed was reduced by more than 50% when using the recirculation filter.

- **Minimize interior dust sources.** Good housekeeping practices are needed to keep enclosure interiors clean, which eliminates inside dust sources. One field study showed a significant increase in dust levels (0.03 to 0.26 mg/m^3) when a floor heater was used. The fan from the floor heater stirred up dust lying on the cab floor [Cecala et al. 2001; NIOSH 2001b].

- **Keep doors closed.** In a study on an enclosed cab of a surface drill, the operator's dust exposure averaged 0.09 mg/m^3 inside the cab with the door closed and 0.81 mg/m^3 when the door was briefly opened to add drill steels [Cecala et al. 2007b]. Although this procedure was performed after drilling stopped and the visible dust dissipated, it nevertheless produced a ninefold increase in dust concentrations inside the cab each time a drill steel was added.

The above research findings also suggest that the use of a one-directional airflow pattern could be beneficial. In most systems, both the intake and discharge for the recirculation air are located in the roof. This could cause a portion of the air to short-circuit without penetrating deeply into the cab. Also, as cab air is drawn into the ventilation system at the roof, dust generated in lower portions of the cab may be pulled through the breathing zone of the worker. In a one-directional design, recirculated air is drawn from the bottom of the enclosure and away from the worker's breathing zone. Figure 4-10 shows this one-directional airflow pattern.

For some cases in operator booths and control rooms, stand-alone table-top air purifier units containing a high efficiency particulate air (HEPA) filter have been installed and shown to improve the air quality. Obviously, these table-top systems are small, portable units available at a fraction of the costs of permanent systems. These systems can be effective if they are sized to handle the volumetric capacity of the booth or control room and if the filters are replaced when necessary [Logson 1998]. Obviously, one shortcoming with this type of system is that they do not provide any pressurization to keep dust from leaking into the booth or room.

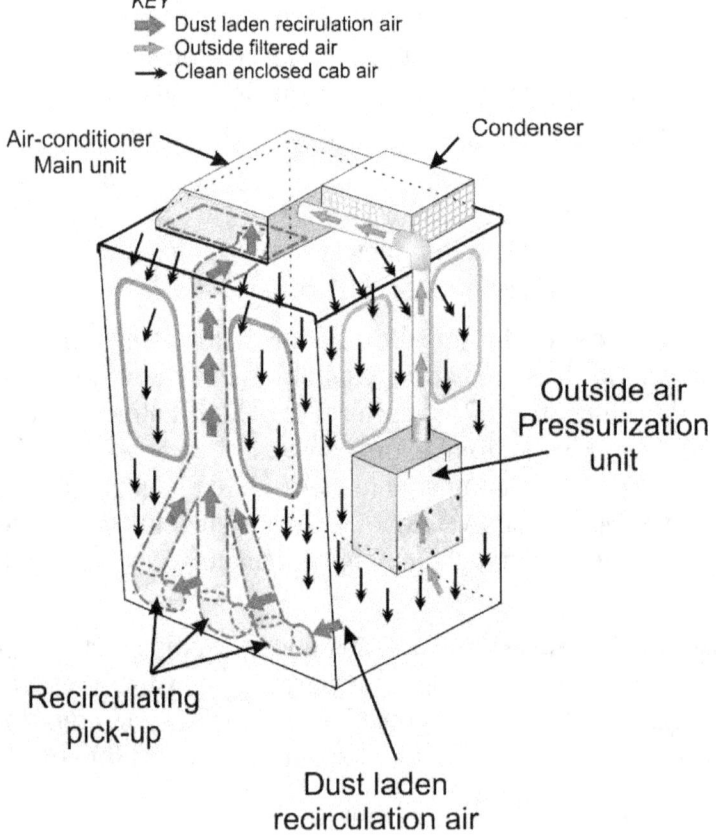

Figure 4-10. Airflow pattern for one-directional filtration system for an enclosed cab.

SCREENING

Screens are the most common device used at mineral processing operations to separate dry ore material into different size ranges, normally measured in mesh sizes. By having the ore processed on different mesh-sized screens, an array of various-sized materials can be produced. Screen sizes for mineral processing operations range from large openings that can be inches in size all the way down to 400 mesh, which has a 35-micron cut point. The amount of dust generated during screening is dependent on the ore type, the particle size, the moisture content, and the type of equipment. Normally, screening finer-sized ore material produces more dust.

Screening has been performed at mineral processing operations for many years and has been perfected by most screen manufacturers. New screening equipment units today are well-sealed units which liberate very little dust into the work environment when they are properly maintained and operated (Figure 4-11). When processing any ore that has significant silica content, screens should be tied into an LEV dust collector system to help keep the process under negative pressure and capture respirable-sized particles liberated within the unit. If visible dust is seen leaking from a screening unit, or if product is visible on the ground underneath a screen, this indicates that a problem exists and needs to be corrected. The first course of action is to determine if a seal or part has worn and is leaking product. In addition, the LEV exhaust port should be examined to ensure it is operating properly and is exhausting the screening unit at the required airflow and pressure.

Another problem occurs when a screen cover is lifted during operation to clean the screen surface. This basically renders the LEV system ineffective because the increased area with the cover lifted is so great that the screen area is not able to be kept under negative pressure. This is a known problem throughout the industry. The only safety option currently available to workers when performing this task is to wear fit-tested personal protective equipment (PPE) that is rated for the levels and type of dust being processed.

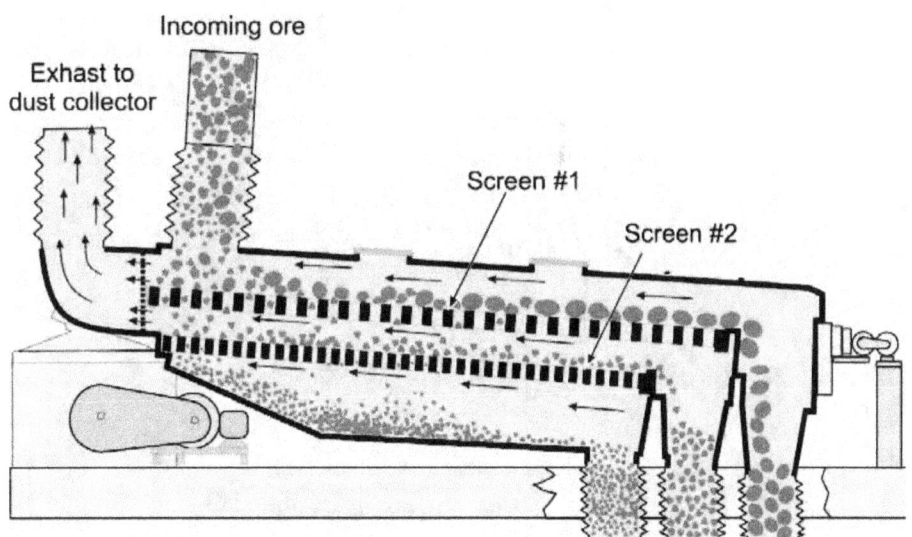

Figure 4-11. Screening unit with LEV system.

PACKAGING/BAGGING PRODUCT FOR SHIPMENT

Ore is processed so that it can be packaged and sold to customers. Mineral processing operations package their product in a wide spectrum that includes 50- and 100-pound bags as well as bulk loading into railcars and trailer trucks. This section will provide some recommended methods to control respirable dust while loading ore into each of the different packaging containers to be delivered to the customer.

Bag Filling Machines Packaging 50- and 100-Pound Bags

- **Dual bag nozzle system.** The dual bag nozzle system was designed to reduce dust from three major sources during bagging [USBM 1984a, b, 1986c; Cecala and Muldoon 1985; Cecala and Thimons 1989]. The system is composed of an improved bag clamp designed to reduce the amount of product blowback during bag filling (Figure 4-12). The clamp reduces blowback by making direct contact with approximately 80% of the fill nozzle. The system incorporates a dual nozzle system, which is a nozzle within a nozzle. The inner nozzle is the normal fill nozzle. The outer nozzle incorporates an air exhaust system, which exhausts excess pressure from the bag when it has finished filling. The exhaust system is powered by an eductor, which uses the venturi effect to exhaust the bag at approximately 50 cfm. Depressurizing the bag reduces the "rooster-tail" of product which spews from the bag when it is ejected from the fill station. By reducing the blowback and rooster tail, the amount of product and dust on the outside of the bag are minimized, which also reduces the dust liberated during this process. During a field analysis on this dual bag nozzle system, the bag operator's dust exposure was reduced by 83% [USBM 1984a].

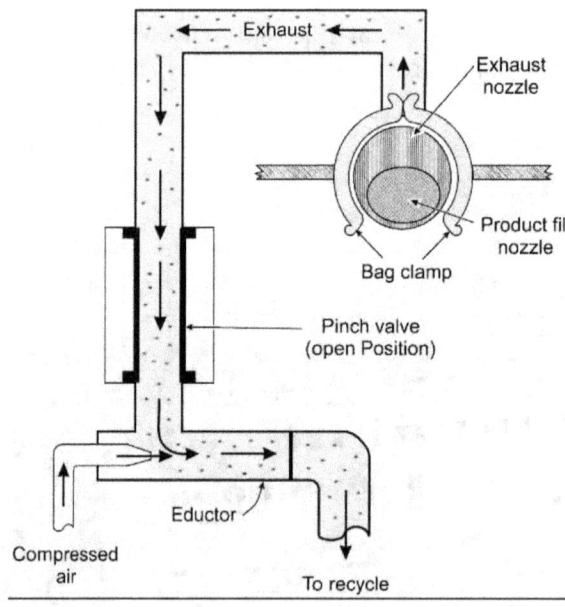

Figure 4-12. Dual bag nozzle design.

- **Overhead air supply island system (OASIS).** The OASIS is used to reduce bag operators' dust exposures while performing the bag loading processes. Normally, the bag operator sits while loading bags, and the OASIS would be installed over his/her position to provide a clean envelope of filtered air down over the worker. (For more information, see the "Background Issues" section of this chapter.)

- **LEV.** A local exhaust ventilation system is very effective at minimizing the dust exposure to the bag operator while loading bags of product. When designing these systems, the dust generated during bag loading must be pulled down toward the floor to prevent the dust from passing over the worker's breathing zone as it is captured by the exhaust ventilation system.

- **Automated systems.** A number of manufacturers have been making advances in automated equipment that mechanically takes 50- or 100-pound bags, loads them onto fill spouts, then ejects the bags once they have reached the desired bagging weight. Normally, the bag loading area is sealed with plastic stripping to isolate it. This area is connected to an LEV system and kept under slight negative pressure to capture and remove respirable-sized dust particles liberated during the bag loading process.

Bag Conveying

- **Bag and belt cleaning devices.** These automated systems use a combination of mechanical devices such as brushes and air sprays to clean dust and product from the outside of the bags as they are being conveyed from the loading station to the stacking area [USBM 1995; Cecala et al. 1997]. It is recommended that bag cleaning be performed in an enclosed system so that all the dust removed is contained and exhausted to an LEV system. By having a hopper located under this cleaning unit, product removed from the bags can be recycled back into the process (Figure 4-13).

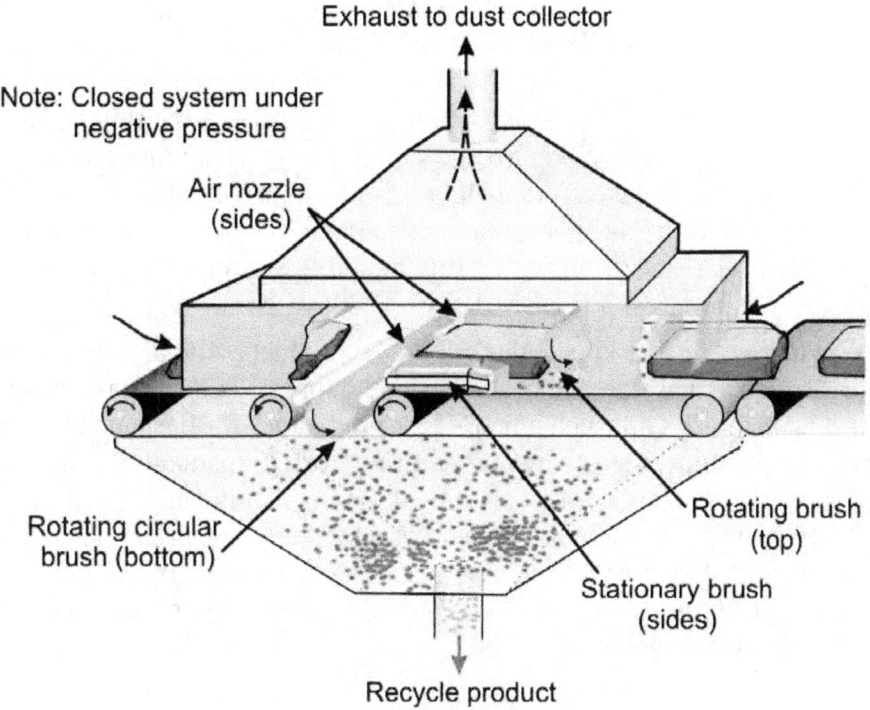

Figure 4-13. Bag and belt cleaner device.

- **Bag valve type.** A study was performed that evaluated the dust liberation differences in relation to five commercial bag valves: (1) standard paper, (2) polyethylene, (3) extended polyethylene, (4) double trap, and (5) foam. This study showed that the bag operator's and bag stacker's respirable dust exposures were reduced by 62% and 66%, respectively, when using the extended polyethylene valve, as compared to the standard paper valve [USBM 1986a; Cecala and Muldoon 1986]. The extended polyethylene valve was a plastic liner 4½ inches inside a standard paper valve. The plastic was then extended for another 1½ inches. It appears that this plastic section allowed the bag valve to seal very effectively to keep product and dust from escaping from this valve. This study was performed two decades ago, and during the time of the study the additional cost was less than one cent per bag, which was very cost-effective when considering the significant reductions measured to both the bag operator and the bag stacker.

Pallet Loading

- **LEV.** When the pallet loading process is manually performed by bag stackers, an LEV system should be incorporated to capture and remove the dust generated during this process. When designing these systems, the dust generated during bag loading must be pulled down toward the floor to prevent the dust from passing over the worker's breathing zone as it is captured by the exhaust ventilation system.

- **OASIS.** The OASIS is also applicable to the pallet loading process to reduce the dust exposure to the bag stacker(s). The OASIS should be installed over the work station and provide a clean envelope of filtered air down over the bag stacker(s).

- **Semiautomated system.** Semiautomated systems use workers in conjunction with an automated system to perform the bag stacking process. This can include a vast array of different setups and types of systems. In one case, a worker performs the bag stacking task manually but is assisted by using a hydraulic lift table. This lift table allows the height for stacking the bags to remain constant throughout the entire pallet loading cycle. The bag loading height is set to approximately knuckle-high for the worker, which is the most ergonomic loading height. A push-pull ventilation system is used on either side of this pallet (Figure 4-14) to capture the dust liberated during bag stacking processing [USBM 1988a, 1989].

  In other cases, the workers slide the bags of product on an air table one layer at a time, but the actual stacking of the bags onto the pallet is performed automatically. Since back injuries are such a major lost-time injury for bag stackers, this design significantly reduces stress by not requiring them to manually lift any bags of product. One problem with an air slide device is that it can cause dust to be blown from the bags of product into the worker's breathing zone. In this case, either an exhaust hood should be placed over the air slide area that is tied into an LEV system, or an OASIS-type system should be placed over the worker [Cecala et al. 2000; NIOSH 2001a].

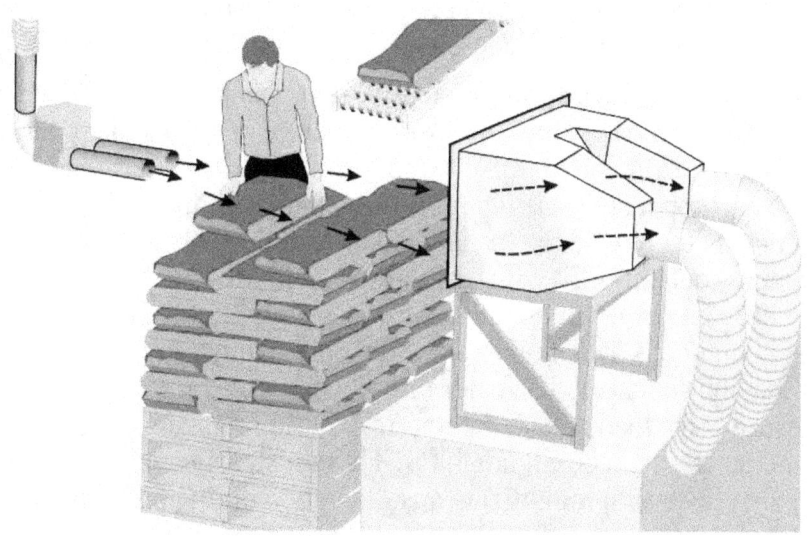

Figure 4-14. Semiautomated pallet loading system using push-pull ventilation.

- **Automated systems.** Significant advances have been made by different manufacturing companies to design systems that either mechanically or robotically performs the bag stacking process, and these systems are being used at mineral processing plants. Automated systems reduce fatigue (back and other body parts) of the worker performing the task manually, as well as, the respirable dust exposure to the worker [Cecala and Covelli 1990]. When using these automated systems, the amount of respirable dust liberated still needs to be evaluated and controlled to ensure that it does not remain on the product bags and does not escape to the surrounding environment. This is normally performed by enclosing the area and using an LEV system to exhaust the respirable dust from the area.

One-Ton Bulk Bags

One-ton bulk bags of product material have become more popular over recent years because they are more cost effective than the 50- or 100-pound bags. The most effective method to control the dust liberated during the one-ton bulk bagging is to isolate this area from the rest of the plant. A worker enters this area and manually attaches an empty bag to the loading device. The worker then leaves the area and once outside, remotely activates a start button to begin bag filling. The area is exhausted and under negative pressure by virtue of being tied into a LEV system. Once bag loading is completed, the worker reenters the area and removes the bag for shipping and begins the process again.

Bulk Loading

- **Enclosed cabs.** Frequently, bulk loading is performed using a front-end loader to load open-container trailer trucks. This is normally performed outdoors with coarser product material, which does not generate a substantial amount of respirable dust. Nevertheless, the front-end loader should have a structurally competent enclosed cab with an effective filtration and pressurization system. (See the "Operator Booths/Control Rooms/Enclosed Cabs" section.)

- **Telescoping loadout spout.** Many different manufacturers produce telescoping loadout spouts, which minimize the amount of dust generated during the bulk loading process. Allowing ore to fall or drop significant distances creates more dust than when the drop distance is minimized. To minimize the drop distance, telescoping loadout spouts extend the loading location so that it is directly above the ore. As the shipping containers fill with ore, the spout continually repositions upward until the container is completely loaded. Another feature of this device that further minimizes dust liberation is a dual tube arrangement. The inner tube delivers the ore product to the loading container. The outer tube is tied into an LEV system to create a negative pressure at the loading area, therefore exhausting the respirable-sized particles to a dust collector system (Figure 4-15). These systems have been used for many years and are effective at minimizing dust liberation during bulk loading.

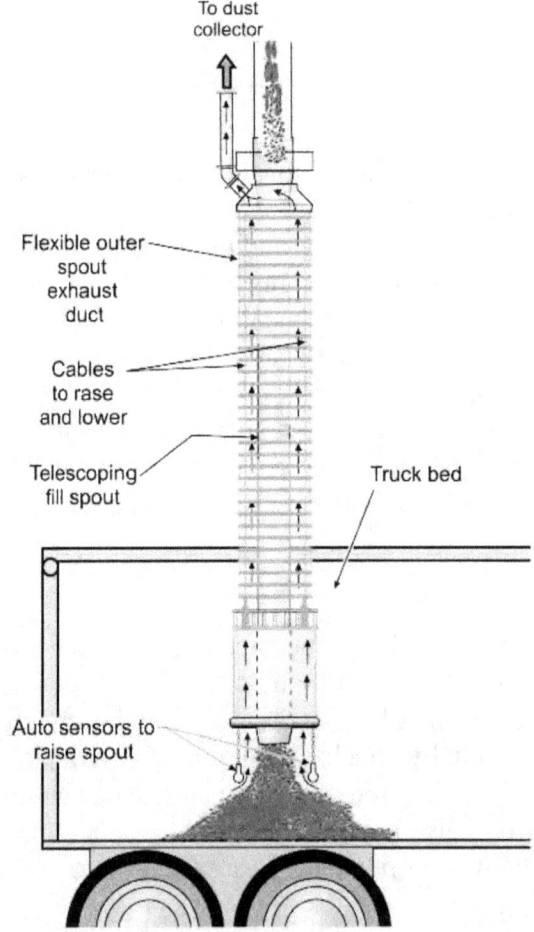

Figure 4-15. Telescoping bulk loading spout with an exhaust system.

CLOTHES CLEANING SYSTEMS

One significant area of respirable dust exposure to workers at mineral processing operations is from contaminated work clothing. For the mineral processing industry, a U.S. Bureau of Mines report documented two cases where a tenfold increase in a worker's respirable dust exposures came from dusty work clothes [USBM 1986b]. This study indicated that respirable dust levels liberated from the soiled clothes were elevated to the extent that workers could be over their permissible exposure limit (PEL) in less than two hours. As the individuals performed their work duties, dust was continually emitted from their clothing, and the only way to eliminate this dust source was to clean or change their work clothing. Although disposable coveralls have been in use for many years, as well as some new and improved clothing material less susceptible to dust capture, the vast majority of workers continue to wear clothing similar to what they have worn for years. In addition, dirty work clothing, if it is not cleaned or changed at the end of the shift, can also contaminate personal vehicles and expose family members [Langenhove and Hertleer 2004; Hartsky et al. 2000; Salusbury 2004].

The MSHA-approved method of cleaning work clothing involves a filtered vacuuming system, which is both difficult and time-consuming for the worker. Because of this, workers sometimes use a single compressed-air hose to blow dust from their clothing, even though this is not an approved practice. This technique creates a significant dust cloud which increases the worker's respirable dust exposure and contaminates the work area.

To address these problems, NIOSH and Unimin Corporation developed a clothes cleaning system that is able to quickly, effectively, and safely remove dust from a worker's clothing without exposure to the worker, the work environment, or coworkers during the cleaning process [NIOSH 2005]. This system has been shown to be significantly more effective than the vacuuming or single air hose technique, while being performed in a fraction of the time.

The clothes cleaning system consists of four major components. Figure 4-16 shows the various components of the clothes cleaning system.

- **Cleaning booth.** The cleaning booth used for testing has a base dimension of 48 inches by 42 inches and provides a safe and controlled area to perform the clothes cleaning process. All intake air enters the cleaning booth through a 24-inch cutout on the roof. The air flows directly down and over the worker in the booth before flowing through expanded metal grating on the floor and exiting through a return air plenum at the bottom and back of the booth. All dust and product cleaned from the worker's clothing is contained within the booth and then exits via the exhaust ventilation system.
- **Air spray manifold.** The air spray manifold is composed of 26 spray nozzles, spaced 2 inches apart, to remove the dust and product from the worker's clothing [Pollock et al. 2005]. These spray nozzles are regulated to limit the operating pressure to a maximum of 30 psi. The top 25 spray nozzles are flat-fan air nozzles and are used to clean the clothing. The bottom nozzle is a circular design and is used for cleaning the individual's work boots.
- **Air reservoir.** The air reservoir supplies the required air volume necessary for the air nozzles used in the air spray manifold. A 240-gallon unit reservoir is recommended for the system because it allows for multiple cleanings to be performed one after another. The air reservoir is located next to the cleaning booth and hard-piped to the air spray manifold located inside the booth.

- **Exhaust ventilation system.** Exhaust ventilation is used to keep the cleaning booth under negative pressure throughout the entire clothes cleaning process. This exhaust ventilation system can be tied into a local exhaust ventilation system or directed outside the plant and exhausted from an elevated stack. Testing on this system has verified that an exhaust volume of 2,000 cfm is required to maintain a negative pressure throughout the entire clothes cleaning cycle [Pollock et al. 2006]. This exhaust ventilation system should only be operated when a worker enters the booth to perform a clothes cleaning cycle and can be de-activated once the process is completed and the worker exits the booth.

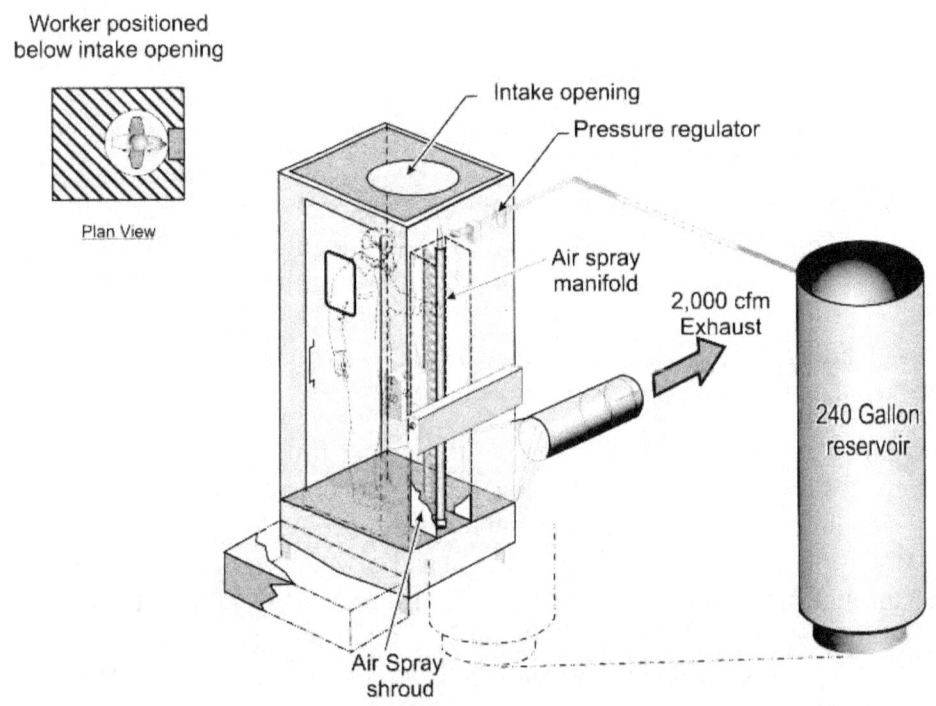

Figure 4-16. Clothes cleaning system design.

During the development of the clothes cleaning system, a matrix of tests was performed to evaluate the effectiveness of this technique in comparison to that of the HEPA vacuuming and the single compressed-air hose approach. For this testing, both 100% cotton and cotton/polyester blend coveralls were soiled with dust before a worker entered the cleaning booth. The new clothes cleaning technique was proven to be 40.8% and 50.6% more effective than the vacuuming and single compressed-air hose technique, respectively [Cecala et al. 2007a]. The clothes cleaning system was also superior in its ability to uniformly remove dust from all areas of the worker's clothing. Another major benefit was that the complete cleaning process was performed in a fraction of the time. The average cleaning times were 317 seconds for vacuuming, 178 seconds for the air hose, and 18 seconds for the clothes cleaning system. This test also indicated that polyester/cotton blend coveralls were cleaned more effectively than the 100% cotton coveralls. Figure 4-17 shows the effectiveness of the technique before and after a worker performed the clothes cleaning process while weaing polyester/cotton blend coveralls.

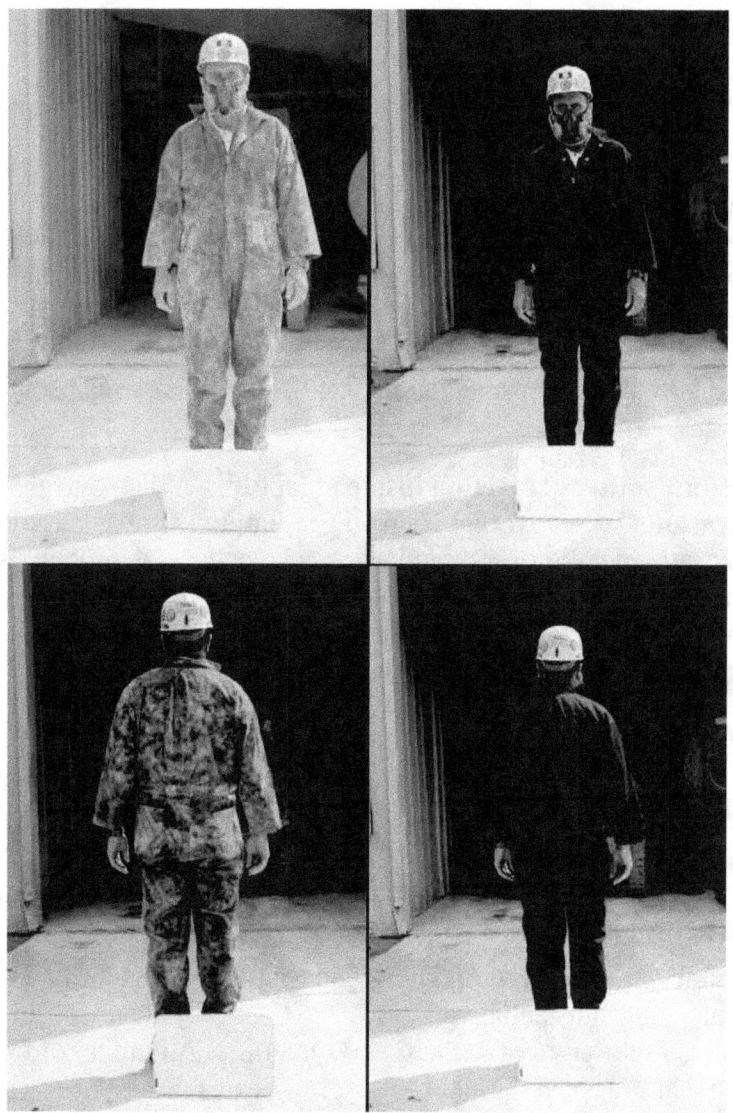

Figure 4-17. Test subject before and after using the clothes cleaning booth.

All workers performing the cleaning process are required to wear a half-mask, fit-tested respirator with N100 filters, hearing protection, and eye protection. To perform the clothes cleaning process, the worker enters the booth wearing his/her PPE, pushes the start button, slowly spins in front of the air spray manifold (18 seconds), and exits the booth with clean clothing. This clothes cleaning system provides a quick and effective method for workers to clean dusty clothes during the workday without risk to the worker, coworkers, or the work environment [Cecala et al. 2005b]. MSHA recognized the benefits of this system and has approved Petitions for Modification so that the system can be used in place of HEPA vacuuming.

BACKGROUND ISSUES

This area of the report will focus on four different background issues: secondary dust sources, open structure design, overhead air supply island system, and housekeeping practices.

Secondary dust sources When a worker is overexposed to respirable dust, most often the assumption is that the dust exposure came from the worker's primary job function. A study was performed that documented a number of cases where this was not the case [Cecala and Thimons 1987]. When a worker obtains a high respirable dust measurement, the correct course of action is to evaluate the worker's job function to determine the dust sources that contributed to this exposure and the magnitude of the exposure from each of these sources. Sometimes a secondary or background dust source can be the major contributor to the worker's overall exposure. Controlling these less obvious dust sources can have a major impact on bringing levels back to acceptable concentrations.

The following examples will demonstrate the impact that secondary dust sources can have on a worker's respirable dust exposure:

- **Outside dust sources traveling inside structures.** When outside dust sources travel inside structures, every worker inside the structure is impacted. Most bagging operations at mineral processing plants use an exhaust ventilation system to draw the dust generated from the bagging process down into the fill hopper. It is important that the air being drawn into this exhaust ventilation system, commonly called make-up air, be clean air. At one operation, the make-up air was drawn directly from the bulk loading area outside the mill. The dust generated from this bulk loading process traveled through an open door into the mill, substantially contaminating the workers inside the mill. During periods when bulk loading was not performed, the bag operator's dust exposure was 0.17 mg/m^3. As trucks were loaded at the bulk loading area, the bag operator's exposure increased to 0.42 mg/m^3 due to this contaminated air [USBM 1986b]. If outside air is used as make-up air, it must be from a location where the air is not contaminated.

- **Performance of job function.** During an evaluation of a dust control system at one processing plant, substantial variations existed in the dust exposures of two different workers due to differences in their work practices. A number of factors were identified that impacted these differences.

    One factor was the amount of time the bag operator allowed the bag to remain on the fill spout before removing it. If the bag remained on the fill spout for a few seconds after it was filled, there was less dust generated from the rooster tail of product that spewed from the bag valve and fill nozzle as the bag was removed. A second factor was the extent to which the bag valve was sealed by the bag operator. One operator did not pay attention to where he grasped the bag as he lifted it from the fill spout to transfer it onto a conveyor belt. A second operator grasped the bag at the fill spout and crimped it closed as he placed the bag on the conveyor. This substantially lowered the amount of product that spewed from the bag as it was placed on the conveyor. A third factor impacting the operator's dust exposure was the general manner in which the operator removed the bag from the bag spout and placed it on the conveyor. More dust was generated when this was done in a forceful manner, as compared to a more continuous and gentle fashion [Cecala and Thimons 1993].

A number of modifications were tested to lower these workers' respirable dust exposures. However, regardless of the effectiveness of the dust control system, the worker who performed his work duties in a rough and careless manner had approximately a 70% higher respirable dust exposure when compared to his coworker who performed the tasks in a conscientious and gentler manner.

- **Broken bags of product.** In most cases, bag breakage occurs because of flaws in the bags delivered from the bag manufacturer. At one particular operation, a bag operator's dust exposure went from 0.07 mg/m^3 before the bag break to 0.48 mg/m^3 afterwards. Although the bag broke during the conveying process and not directly in front of the worker, the dust substantially contaminated the surrounding mill air, which flowed over the bag operator. Once again, this occurred because the exhaust ventilation system in the bag loading area created a negative pressure that draws background air from the mill.

For mineral processing operations to keep workers at acceptable dust levels, management must be aware of the various dust contamination sources and methods to reduce these sources. The substantial effects of the various secondary dust sources should be recognized, identified, and controlled in an effort to minimize workers' dust exposures.

Open-structure design

Many different types of structures and materials have been used to build mineral processing facilities through the years. Although structure type and building material were not viewed as significant factors affecting the health of employees in these facilities when they were built, a recent study was performed that compared dust levels with three different building types: masonry, an open-structure design, and a steel-sided design [Cecala et al. 2007c]. Respirable dust measurements were taken within these structures to evaluate and compare levels. When considering the data, the most effective structural design of these three building types from a dust standpoint was an open-structure design. Respirable dust concentrations were significantly lower in the open structure because the natural environment acts as the best source of ventilation to dilute and carry away dust generated and liberated within the structure. In this study, when production levels were normalized, respirable dust levels at the open structure were more than four times lower than at the masonry structure and over 1,000 times lower than at the steel-sided structure [Cecala et al. 2006].

From a federal standard basis, the only consideration for an open-structure design would be the Environmental Protection Agency's opacity dust measurement, which is a qualitative measurement taken by a federal regulator based on a visible dust plume. If any plumes are visible, source dust controls must be implemented to address the problem.

Figure 4-18 shows a conceptual drawing of a typical walled processing facility, then an identically sized facility with an open-structure design and a roof. Obviously, a roof would provide a little more protection from the natural elements than a totally open design. When building new facilities, the open-structure design is more cost-effective because there are lower material and construction costs involved. Some companies may also want to consider modifying their existing structures with a more open design to further reduce dust levels. If an open-structure design is considered for an operation, a number of issues need to be addressed:

- Safety railings, guards, and/or chain-link wall containments must be installed to minimize the potential for any personnel or objects falling from the structure.
- Equipment and personnel must be protected from environmental elements such as rain, snow, sleet, and hail. One possibility to minimize this concern would be to design a structure with a sufficient overhang.
- An open-structure design must be considered a secondary design. The first approach to lower dust exposures in any structure is to have an effective primary dust control plan that captures major dust sources at their point of origin, before they are allowed to liberate into the plant and contaminate workers.

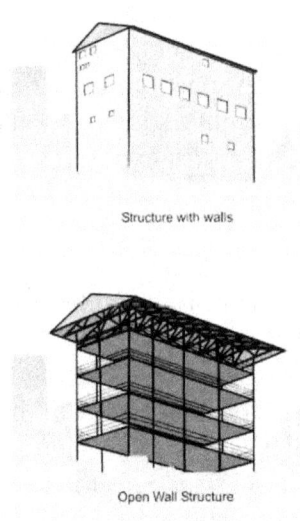

Figure 4-18. Drawing of a conventional and an open structure with a protective overhang.

Overhead air supply island system

A successful control technique to reduce respirable dust exposures at mineral processing operations when workers are at stationary positions is with Overhead Air Supply Island System (OASIS). The OASIS air cleaning device is suspended over a worker and provides a flow of filtered air over the work station. Mill air is drawn into the system and passed through a primary cartridge filter. This primary filter is self cleaning, automatically using the reverse pulse technique when excessive filter pressure is sensed. The air can then pass through a heating or cooling chamber, which is optional depending on the mill air temperature, and from there into a distribution manifold, which also serves as a secondary filter (Figure 4-19). The resulting filtered air flows down over the worker at an average velocity of approximately 400 fpm, which restricts mill air from entering the clean air zone.

During development, OASIS systems were installed over bag operators at two different mineral processing operations. Testing at these two operations showed 98% and 82% reductions in the bag operators' respirable dust exposures. An additional benefit with this system was a 12% reduction in general dust levels throughout the mill building as a result of the OASIS cleaning the mill air [Volkwein et al. 1988].

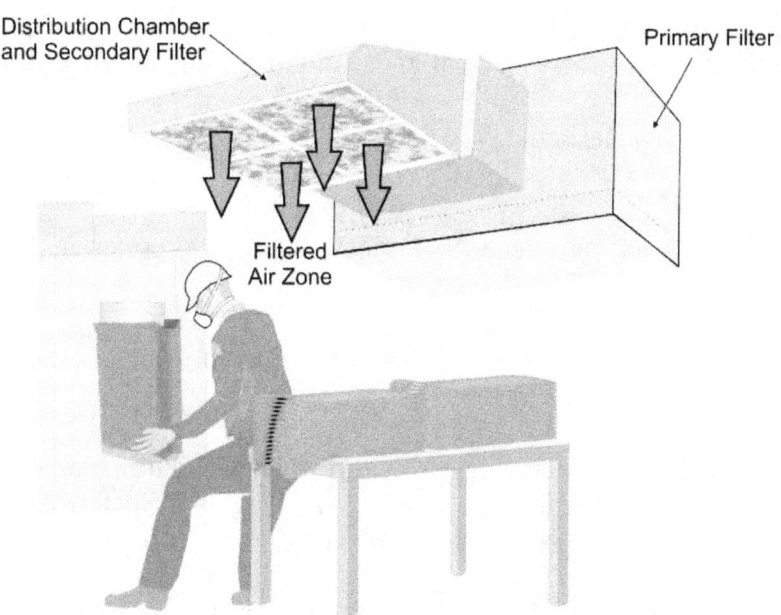

Figure 4-19 - Overhead air supply island system.

Housekeeping practices

Although good housekeeping practices seem to be a minor or common-sense issue, it can be a significant factor in a worker's respirable dust exposure at mineral processing operations. When housekeeping is performed properly and on a scheduled time frame, it can play a key factor in minimizing respirable dust exposure to workers at processing plants. When it is not performed, or performed improperly, it can have just the opposite effect. One example of this was documented when a worker was dry sweeping the floor with a push broom at a mineral processing operation. In this instance, the exposure of a coworker located one floor up from the person sweeping the floors increased from 0.03 mg/m^3 before sweeping occurred to 0.17 mg/m^3 during and immediately after the occurrence. Dry sweeping is an unacceptable method of cleaning because of the dust it liberates into the work environment [USBM 1986b].

The most effective method of housekeeping is to wash down the plant with water on a regular basis. For the ideal system, floor drains and floors that are sloped correctly toward the drains should be incorporated into the structure's design during construction. Housekeeping should be performed at the end of each shift so that workers coming on to the next shift start with a clean work environment.

A final component of effective housekeeping is proper upkeep and maintenance of plant equipment and processes. When product is observed building up on the floors of the plant, it indicates that some function or process is leaking ore. In some cases, visible dust can be seen leaking from holes or damaged equipment and this must be quickly corrected to minimize dust leakage.

REFERENCES

ACGIH [2007]. Industrial ventilation handbook: a manual of recommended practice for design, 26th edition. Cincinnati, OH: American Conference of Governmental Industrial Hygienists.

Bartell W, Jett B [2005]. The technology of spraying for dust suppression. Cement Americas *May/June*:32–37.

Bresee R [2008]. Vice President Technical Services, Unimin Corporation, personal correspondence.

Cecala AB [1998]. Supplementing your dust control equipment with whole-plant ventilation. Powder Bulk Eng *12*(1):19–32.

Cecala AB, Covelli A [1990]. Automation to control silica dust during pallet loading process. Paper presented at the SME Annual Meeting. Salt Lake City, UT: Society for Mining, Metallurgy, and Exploration, Inc., Preprint 90-28, 5 p.

Cecala AB, Mucha R [1991]. General ventilation reduces mill dust concentrations. Pit Quarry *84*(1):48–53.

Cecala AB, Muldoon T [1985]. New bag nozzle system reduces dust generated during bag filling. Rock Prod *July*:32–33.

Cecala AB, Muldoon T [1986]. Closing the door on dust—dust exposure of bag operator and stackers compared for commercial bag valves. Pit Quarry *78*(11):36–37.

Cecala AB, Thimons ED [1987]. Significant dust exposures from background sources. Pit Quarry *79*(12):46–51.

Cecala AB, Thimons ED [1989]. New dual bag nozzle system. Ceram Eng Sci Proc *10*(1-2):36–41.

Cecala AB, Thimons ED [1993]. Tips for reducing dust from secondary sources during bagging. Powder Bulk Eng *7*(5):77–84.

Cecala AB, Thimons ED [1997]. Methods to lower dust exposures at mineral processing operations. In: Proceedings of the National Stone Association Meeting—a compliance for the 21st century. St. Louis, MO: National Stone Association, October, pp. 231–249.

Cecala AB, Klinowski GW, Thimons Ed [1995]. Reducing respirable dust concentrations at mineral processing facilities using total mill ventilation system. Min Eng *47*(6):575–576.

Cecala AB, Daniel JH, Thimons ED [1996]. Methods to lower dust exposure at mineral processing operations. Appl Occup Environ Hyg J *11*(7):854–859.

Cecala AB, Timko RJ, Prokop AD [1997]. Bag and belt cleaner reduces employee dust exposure. Rock Prod *100*(3):41–43.

Cecala AB, Zimmer JA, Smith B, Viles S [2000]. Improved dust control for bag handlers. Rock Prod *103*(4):46–49.

Cecala AB, Organiscak JA, Heitbrink WA [2001]. Dust underfoot—enclosed cab floor heaters can significantly increase operator's respirable dust exposure. Rock Prod *104*(4):39–44.

Cecala AB, Organiscak JA, Heitbrink WA, Zimmer JZ, Fisher T, Gresh RE, Ashley JD II [2004]. Reducing enclosed cab drill operator's respirable dust exposure at surface coal operations with a retrofitted filtration and pressurization system. In: SME Transactions 2003, Vol. 314. Littleton, CO: Society for Mining, Metallurgy and Exploration, Inc., pp. 31–36.

Cecala AB, Organiscak JA, Zimmer JA, Heitbrink WA, Moyer ES, Schmitz M, Ahrenholtz E, Coppock CC, Andrews EH [2005a]. Reducing enclosed cab drill operator's respirable dust exposure with effective filtration and pressurization techniques. J Occup Environ Hyg *2*:54–63.

Cecala AB, Zimmer JA, Colinet JF, Timko RJ, Chekan GJ, Pollock DE [2005b]. Control technology using ventilation to reduce respirable dust exposure at U.S. metal/nonmetal mining operations. In: Proceedings of the 8th International Mine Ventilation Congress. Brisbane, Australia: International Mine Ventilation Congress, July 6.

Cecala AB, Rider JP, Zimmer JA, Timko RJ [2006]. Lower respirable dust and noise exposure with an open structure design. Cincinnati, OH: U.S. Department of Health and Human Services, Centers for Disease Control and Prevention, National Institute for Occupational Safety and Health, DHHS (NIOSH) Publication No. 2007-101.

Cecala AB, O'Brien AD, Pollock DE, Zimmer JA, Howell JL, McWilliams LJ [2007a]. Reducing dust exposure of workers using an improved clothes cleaning process. Int J Min Res Eng *2*(2):73–94.

Cecala AB, Organiscak JA, Zimmer JA, Moredock D, Hillis M [2007b]. Closing the door to dust when adding drill steels. Rock Prod *Oct.*29–32.

Cecala AB, Rider JP, Zimmer JA, Timko RJ [2007c]. Dial down dust and noise exposure. Aggregates Manag *12*(7):50–53.

Chekan GJ, Colinet JF [2003]. Retrofit options for better dust control—cab filtration, pressurization systems prove effective in reducing silica dust exposures in older trucks. Aggregates Manag *8*(9):9–12.

Courtney WG [1983]. Single spray reduces dust 90%. Coal Min Proc *June*:75–77.

Ford VHW [1973]. Bottom belt sprays as a method of dust control on conveyors. Min Tech (UK) *55*(635):387–391.

Hartsky MA, Reed KL, Warheit DB [2000]. Assessments of the barrier effectiveness of protective clothing fabrics to aerosols of chrysotile asbestos fibers. In: Nelson C, Henry N, eds. Performance of protective clothing: issues and priorities for the 21st century. Vol. 7. West Conshohocken, PA: American Society for Testing and Materials, pp. 141–154.

Langenhove LV, Hertleer C [2004]. Smart clothing: a new life. Int J Cloth Sci Tech *16*(1,2):63–72.

Logson R [1998]. Controlling respirable dust in plant control rooms. Stone Rev *14*(6):43–44.

MAC [1980]. Design guidelines for dust control at mine shafts and surface operations, third edition. Ottawa, Ontario, Canada: Mining Association of Canada.

Martin Marietta Corp [1987]. Dust control handbook for minerals processing. Raleigh, NC: Martin Marietta Corporation. USBM contract no. J0235005.

MSA [1978]. Improved dust control at chutes, dumps, transfer points, and crushers in noncoal mining operations. NTIS No. PB 297–422. By Rodger SJ, Rankin RL, Marshall MD. MSA Research Corp., USBM contract no. H0230027.

MSHA [2009]. Number of metal/nonmetal operations in the United States [http://www.msha.gov/ STATS/PART50/WQ/1978/wq78mn01.asp]. Date accessed: September 2009.

NIOSH [2001a]. Hazard control 31: Dust protection for bag stackers. Cincinnati, OH: U.S. Department of Health and Human Services, Centers for Disease Control and Prevention, National Institute for Occupational Safety and Health.

NIOSH [2001b]. Technology news 486: floor heaters can increase operator's dust exposure in enclosed cabs. Cincinnati, OH: U.S. Department of Health and Human Services, Centers for Disease Control and Prevention, National Institute for Occupational Safety and Health.

NIOSH [2003]. Handbook for dust control in mining. By Kissell FN. Cincinnati, OH: U.S. Department of Health and Human Services, Centers for Disease Control and Prevention, National Institute for Occupational Safety and Health NIOSH IC 9465.

NIOSH [2005]. Technology news 509: a new method to clean dust from soiled work clothes. U.S. Department of Health and Human Services, Centers for Disease Control and Prevention, National Institute for Occupational Safety and Health, DHHS (NIOSH) Publication no. 2005-136.

NIOSH [2007]. Technology news 528, recirculation filter is key to improving dust control in enclosed cabs. By Organiscak JA, Cecala AB. Cincinnati, OH: U.S. Department of Health and Human Services, Centers for Disease Control and Prevention, National Institute for Occupational Safety and Health, DHHS (NIOSH) Publication No. 2008-100.

NIOSH [2008]. Key design factors of enclosed cab dust filtration systems. By: Organiscak JA, Cecala AB. Cincinnati, OH: U.S. Department of Health and Human Services, Centers for disease Control, National Institute for Occupational Safety and Health, DHHS (NIOSH) Publication No. 2009-103.

Organiscak JA, Cecala AB, Thimons ED, Heitbrink WA, Schmitz M, Ahrenholtz E [2004]. NIOSH/Industry collaborative efforts show improved mining equipment cab dust protection. In: SME Transactions 2003, Vol. 314. Littleton, CO: Society for Mining, Metallurgy and Exploration, Inc., pp. 145–152.

Planner JH [1990]. Water as a means of spillage control in coal handling facilities. In: Proceedings of the Coal Handling and Utilization Conference. Sydney, Australia: Institution of Engineers Australia, pp. 264–270.

Pollock DE, Cecala AB, O'Brien AD, Zimmer JA, Howell JL [2005]. Dust off. Rock Prod *March*.

Pollock DE, Cecala AB, Zimmer JA, O'Brien AD, Howell JL [2006]. A new method to clean dust from soiled work clothes. In: Proceedings of the 11th U.S./North American Mine Ventilation Symposium. University Park, PA: Pennsylvania State University, pp. 197–201.

Quilliam JH [1974]. Sources and methods of control of dust. In: The ventilation of south African gold mines. Yeoville, Republic of South Africa: The Mine Ventilation Society of South Africa.

Roberts AW, Ooms M, Bennett D [1987]. Bulk solid conveyor belt interaction in relation to belt cleaning. Bulk Solids Handling *7*(3):355–362.

Salusbury I [2004]. Tailor-made-self-cleaning clothing, chemical warfare suits that trap toxins. Materials World *August*:18–20.

Stahura R, Marti A [1995]. Conveyor trends. World Min Equip *19*(5):12–21.

USBM [1976]. Water spray system for continuous-mining machines. By Divers EF. Washington, DC: U.S. Department of the Interior, U.S. Bureau of Mines, IC 8727.

USBM [1981]. Technology News 20: Improved filtering system for water sprays resists clogging. Pittsburgh, PA: U.S. Department of the Interior, U.S. Bureau of Mines.

USBM [1984a]. New bag nozzle system reduces dust from fluidized air bag machines. By Cecala AB, Volkwein JC, Thimons ED. Washington, DC: U.S. Department of the Interior, U.S. Bureau of Mines, RI 8886.

USBM [1984b]. Technology News 207: New bag nozzle system reduces dust during bagging operation. Pittsburgh, PA: U.S. Department of the Interior, U.S. Bureau of Mines.

USBM [1986a]. Dust reduction capabilities of five commercial bag valves. By Cecala AB, Covelli A, Thimons ED. Washington, DC: U.S. Department of the Interior, U.S. Bureau of Mines, IC 9068.

USBM [1986b]. Impact of background sources on dust exposure of bag machine operator. Cecala AB, Thimons ED. Washington, DC: U.S. Department of the Interior, U.S. Bureau of Mines, IC 9089.

USBM [1986c]. Technology News 240: Reduce dust exposure during bag filling and stacking. Pittsburgh, PA: U.S. Department of the Interior, U.S. Bureau of Mines.

USBM [1988a]. Pallet loading dust control system. By Cecala AB, Covelli A. Washington, DC: U.S. Department of the Interiour, U.S. Bureau of Mines, RI 9197.

USBM [1989]. Technology News 328: New pallet loading system lowers worker's dust exposure and improves bag stacking process. Pittsburgh, PA: U.S. Department of the Interior, U.S. Bureau of Mines.

USBM [1993]. Reducing respirable dust concentrations at mineral processing facility using total mill ventilation system. By Cecala AB, Klinowski GW, Thimons ED. Washington, DC: U.S. Department of the Interior, U.S. Bureau of Mines, RI 9469.

USBM [1995]. Reducing respirable dust levels during bag conveying and stacking using bag and belt cleaner device. By Cecala AB, Timko RJ, Prokop AD. Washington, DC: U.S. Department of the Interior, U.S. Bureau of Mines, RI 9596.

Volkwein JC, Engle MR, Raether TD [1988]. Dust control with clean air from an overhead air supply island (OASIS). Appl Ind Hyg J *3*(8):236–239.

Weakly A [2000]. Controlling dust without using bag houses. Coal Age *November*:24–26.

Yourt GR [1990]. Design principles for dust control at mine crushing and screening operations. Canadian Min J *10*:65–70.

Zimmer W [2003]. Dust and water—do they really clash? Bulk Solids Handling *23*(5):302–307.

CHAPTER 5. CONTROLLING RESPIRABLE SILICA DUST AT SURFACE MINES

By John A. Organiscak

Overexposure to airborne respirable crystalline silica dust (referred to here as "silica dust") can cause silicosis, a serious and potentially fatal respiratory lung disease. Mining continues to have some of the highest incidences of work-related silicosis, with mining machine operators being the occupation most commonly associated with the disease [NIOSH 2003b]. Some of the most severe cases of silicosis have been observed in surface mine rock drillers [NIOSH 1992]. A voluntary surface coal miner lung screening study initiated in Pennsylvania in 1996 indicated that silicosis was directly related to age and years of drilling experience [CDC 2000].

Mine workers continue to be at risk of exposure to excessive levels of silica dust in the United States. The percentage of the Mine Safety and Health Administration (MSHA) dust samples from 2004 to 2008 that exceeded the applicable or reduced respirable dust standard due to the presence of silica were 12% for sand and gravel mines, 13% for stone mines, 18% for nonmetal mines, and 21% for metal operations [MSHA 2009]. At surface mining operations, occupations most frequently exceeding the applicable respirable dust standard are usually operators of mechanized equipment such as drills, bulldozers, scrapers, front-end loaders, haul trucks, and crushers.

This chapter summarizes the current state of the art dust controls for surface mines. Surface mining operations present dynamic and highly variable silica dust sources. Most of the dust generated at surface mines is produced by mobile earth-moving equipment such as drills, bulldozers, trucks, and front-end loaders excavating silica-bearing rock and minerals. Four practical areas of engineering controls designed to mitigate exposure of surface mine workers to all airborne dusts, including silica, are drill dust collection systems, enclosed cab filtration systems, controlling dust on unpaved haulage roads, and controlling dust at the primary hopper dump.

Many surface mine dust control problems can be visually observed and diagnosed. Visible airborne dust emissions generated from a particular surface mine process usually indicate that respirable silica dust can be present and potentially become a worker exposure problem. Visual dust emissions affecting nearby workers indicates that an engineering control is needed or an existing control needs maintenance. Investigating possible causes of visual dust emissions when using an engineering control can often uncover the reason for its poor dust control effectiveness. Frequent visual inspections of engineering control systems can identify needed maintenance to optimize its dust control effectiveness. Area dust sampling should be conducted in conjunction with personal sampling when workers are being overexposed to respirable silica dust that cannot be observed. Area dust sampling locations are usually selected near potential dust sources to examine their contribution to the worker dust exposure problem.

DRILL DUST COLLECTION SYSTEMS

Drill dust is generated by compressed air (bailing airflow) flushing the drill cuttings from the hole being drilled. Because of their ability to be operated in freezing temperatures, dry dust collection systems tend to be the most common type of dust control method incorporated into drilling machines by original equipment manufacturers. The typical dry dust collection system as shown in Figure 5-1 is comprised of a self-cleaning (compressed air back-pulsing of filters) dry dust collector sucking the dusty air from underneath the shrouded drill deck located over the hole. Ninety percent of dust emissions with this type of system are attributed to drill deck shroud leakage, drill stem bushing leakage, and dust collector dump discharge. Wet suppression is another drill dust collection method and involves injecting water into the bailing airflow traveling down the drill stem. The process of the bailing airflow, water droplets, and cuttings mixing together captures the airborne dust as it travels back up the hole. However, wet suppression is infrequently used because of operational problems in cold climates, lack of a readily accessible water supply, and shorter bit life. Studies by the U.S. Bureau of Mines and the National Institute for Occupational Safety and Health (NIOSH) have shown the practical aspects of optimizing these dust collection systems. These are discussed below for each dust collection method.

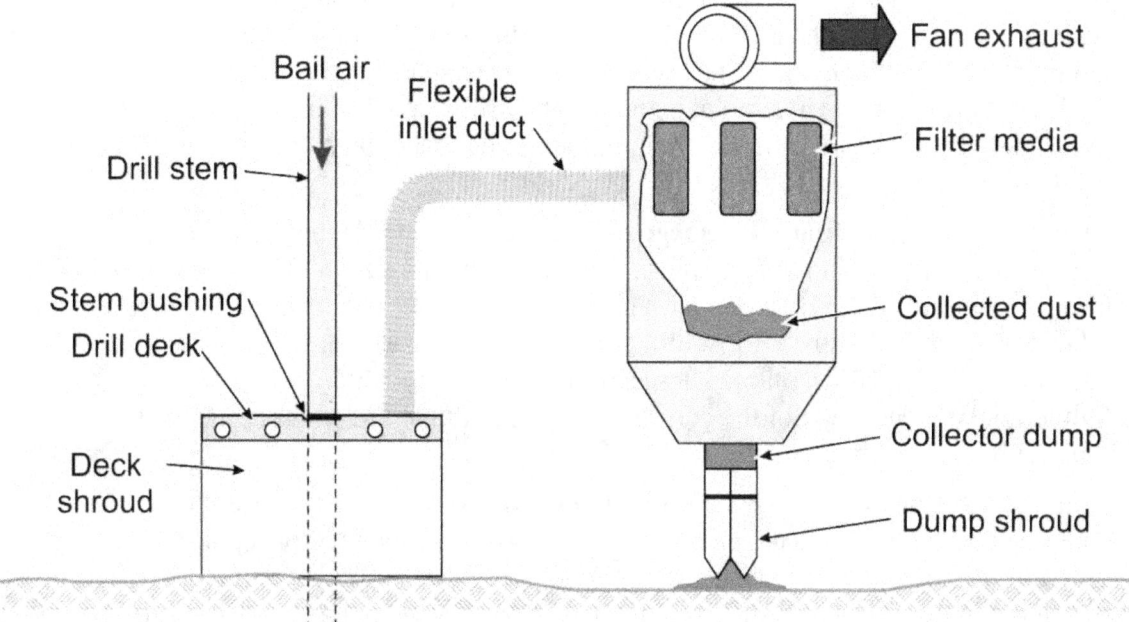

Figure 5-1. Typical dry dust collection system used on surface drills.

Dry Dust Collector System

- **Maintain a tight drill deck shroud enclosure with the ground.** Dust emissions are significantly reduced around the drill deck shroud by maintaining the ground-to-shroud gap height below 8 inches [NIOSH 2005; USBM 1987b]. This can be accomplished by better vertical positioning of the drill table shroud by the operator to minimize the ground-to-shroud gap. Dust levels were significantly reduced from 21.4 mg/m^3 to 2.5 mg/m^3 next to the drill deck shroud when the drill operator changed his drill setup procedure to minimize this gap [Organiscak and Page 1999]. Also, the ground-to-shroud gap can be more tightly closed by using a flexible shroud design that can be mechanically raised and lowered to the ground via cables and hydraulic actuators. An adjustable-height shroud design maintains a better seal with uneven ground and was found to keep dust emissions next to the shroud below 0.5 mg/m^3 at several drill operations [NIOSH 1998, 2005]. Finally, a shroud constructed in sections with vertical gaps along sections or corners can also be a source of shroud leakage. Overlapping sections of shroud material reduces gaps and leakage. One conceptual shroud design for a rectangular drill table is construction with corner sections and overlapping side sections of shroud material [Page and Organiscak 1995].

- **Maintain a collector-to-bailing airflow ratio of at least 3:1.** Dust emissions are significantly decreased around the shroud at or above a 3:1 collector-to-bailing airflow ratio [NIOSH 2005]. Dust collector airflow reductions under the shroud are generally caused by restrictions and/or leakages in the system. Loaded filters and material in the ductwork are likely causes of restrictions, while damaged duct work and holes are likely causes of leakage in the system. Thus, inspection and maintenance of the dust collection system is vital to achieving and maintaining optimal collector operation and airflow.

- **Maintain a good drill stem seal with the drill table.** A rubber drill stem bushing (see Figure 5-1) restricts bailing airflow from blowing dust and cuttings through the drill deck and, therefore, needs to be replaced after mechanical wear. An alternative sealing method involves using a nonmechanical compressed air ring seal manifold under the drill deck. This manifold consists of a donut-shaped pipe with closely spaced holes on the inside perimeter which discharges air jets in a radial pattern at the drill stem. The high-velocity air jets block the gap between the drill stem and deck, reducing respirable dust leakage through the drill deck by 41%–70% [Page 1991].

- **Shroud the collector dump discharge close to the ground.** Dumping dust from the collector discharge several feet above ground level can disperse significant amounts of airborne respirable dust. Dust emission reductions of greater than 63% were measured by the collector discharge dump after installing an extended shroud near ground level (Figure 5-1) [Reed et al. 2004; USBM 1995]. These shrouds can be quickly installed by wrapping brattice cloth around the perimeter of the collector discharge dump and securing it to the discharge dump with hose clamps.

- **Maintain dust collector as specified by manufacturer.** Collector system components should be frequently inspected and damaged components repaired or replaced. A 51% reduction in dust emission was measured at one drill after a broken collector fan belt was replaced, while another drill showed a reduction of 83% after the torn deck shroud was replaced [Organiscak and Page 1999].

Wet Suppression

- **Add small amounts of water into the bailing air until the visible dust cloud has been significantly reduced.** Drill dust emissions are significantly reduced by increasing the water flow rate from 0.2 gpm to 0.6 gpm [USBM 1987b]. A needle valve and water flow meter installed on the water supply line provide adjustable control for wet suppression systems. However, adding excessive water down the hole can cause operational problems with no appreciable improvement in dust control.

- **Minimizing water flow to a rolling cutter bit can increase bit life.** Wet drilling with rolling cutter bits can cause premature bit wear. A drill stem water separator installed upstream of a rolling cutter bit can increase bit life without adverse effects on dust control [Listak and Reed 2007; USBM 1988]. The water separator is a bit stabilizer with an internal cyclonic or impaction water droplet classifier, which removes most of the water from the bailing airflow before it is flushed through the drill bit. The water removed by the internal separator is released through external holes in the bit stabilizer (Figure 5-2).

Figure 5-2. Water separator discharging water before it reaches the drill bit.

ENCLOSED CAB FILTRATION SYSTEMS

Enclosed cab filtration systems are one of the mainstay engineering controls for reducing mobile equipment operators' exposure to airborne dust at surface mines. Enclosed cabs with heating, ventilation, and air conditioning (HVAC) systems are typically integrated into the drills and mobile equipment to protect the operator from the outside environment. Air filtration is often part of the HVAC system as an engineering control for airborne dusts. Surface mining dust surveys conducted by NIOSH on drills and bulldozers have shown that enclosed cabs can effectively control the operator's dust exposure, but cab performance can vary [Organiscak and Page 1999]. The enclosed cab protection factors (outside ÷ inside dust concentration) measured on rotary drills ranged from 2.5 to 84, and those measured on bulldozers ranged from 0 to 45. NIOSH also conducted field studies of upgrading older equipment cabs to improve their dust control effectiveness. These studies involved retrofitting older enclosed cabs with air-conditioning, heating, and air filtration systems to demonstrate the effectiveness of upgrading older mine equipment cabs. During these retrofits, cab enclosure cracks, gaps, or openings were sealed with silicone and closed cell foam tape. Varying degrees of enclosure integrity were achieved. Table 5-1 shows the results in ascending order of performance achieved with these retrofitted installations. Additionally, NIOSH conducted controlled laboratory experiments to examine the key design factors of enclosed cab dust filtration systems. The key performance factors for effective enclosed cab dust filtration systems are summarized below.

Table 5-1. Respirable dust sampling results of enclosed cab field studies.

Cab Evaluation [Ref]	Cab Pressure in w.g.	#Wind Velocity Equivalent mph	Average Inside Cab Dust Level mg/m^3	Average Outside Cab Dust Level mg/m^3	Protection Factor Out/In
Rotary Drill [Organiscak et al. 2004]	None Detected	0	0.08	0.22	2.8
Haul Truck [Chekan and Colinet 2003]	0.01	4.5	0.32	1.01	3.2
Front-end Loader [Organiscak et al. 2004]	0.015	5.6	0.03	0.30	10.0
Rotary Drill [Cecala et al. 2004]	0.20 to 0.40	20.3 to 28.7	0.05	2.80	56.0
Rotary Drill [Cecala et al. 2005]	0.07 to 0.12	12.0 to 15.7	0.07	6.25	89.3

Wind Velocity Equivalent = (4000 $\sqrt{\Delta p_{cab}}$) fpm × 0.11364 mph/fpm @ Standard Air Temp & Pressure.

Key Performance Factors for Enclosed Cab Filtration Systems

- **Ensure good cab enclosure integrity to achieve positive pressurization against wind penetration into the enclosure.** As shown in Table 5-1, significant improvements in cab protection factors were achieved in the field studies when cab pressures exceeded 0.01 inches of water gauge. This corresponded to wind velocity equivalents (an indicator of cab wind velocity resistance) greater than 4.5 miles per hour. The cab enclosures with greater than 0.01 inches of water gauge pressure were of close-fitted construction and their integrity could be readily improved by sealing cab enclosure cracks, gaps, or openings with silicone and closed cell foam tape. The loosely fitted cab construction on one of the drills and the truck were difficult to seal, which limited the amount of cab pressure that could be attained.

- **Use high-efficiency respirable dust filters on the intake air supply into the cab.** Filter efficiency performance specifications used in the field were 95% or greater on respirable-sized dusts [Checkan and Colinet 2003; Cecala et al. 2004, 2005; Organiscak et al. 2004]. Laboratory experiments showed an order of magnitude increase in cab protection factors when using a 99%-efficient filter versus a 38%-efficient filter on respirable-sized particles [NIOSH 2007].

- **Use an efficient respirable dust recirculation filter.** All the cab field demonstrations used recirculation filters that were 95% efficient or better in removing respirable-sized dusts [Checkan and Colinet 2003; Cecala et al. 2004, 2005; Organiscak et al. 2004]. Laboratory experiments showed an order of magnitude increase in cab protection factors when using an 85%- to 94.9%-efficient filter as compared to no recirculation filter [NIOSH 2007]. Laboratory testing also showed that when using a recirculation filter the time for interior cab concentration to decrease and reach stability after the cab door is closed was cut by more than half.

- **Minimize dust sources in the cab.** Use good housekeeping practices and move heater outlets that blow across soiled cab floors. Dust levels were shown to increase from 0.03 to 0.26 mg/m^3 by turning on a floor heater inside the cab [Cecala et al. 2005]. The floor heater was removed and cab heating was discharged down from the ceiling HVAC system, reducing dust entrainment in the cab during colder winter months [Cecala et al. 2005]. Another method of reducing entrainment of dust from a soiled cab floor is placing a gritless (without sand added) sweeping compound on the floor during the working shift. Most commercial sweeping compounds have petroleum-based oils or wax added to the cellulose material. However, people sensitized to petroleum distillates could have allergic reactions to these sweeping compounds if used in enclosed cabs. A few companies offer nonpetroleum-base sweeping compounds that use either a natural oil or chemical additive for dust adhesion [NIOSH 2001]. It is also recommended to cover the floor with rubber matting instead of carpeting for easy cleaning. More frequent cleaning of heavily soiled floors may be a more straightforward alternative than using sweeping compounds to minimize this type of dust entrainment.

- **Keep doors closed during equipment operation.** On one drill operation, the respirable dust concentrations inside the cab averaged 0.09 mg/m^3 with the door closed and averaged 0.81 mg/m^3 when the door was briefly opened to add drill steels [Cecala et al. 2007]. Although this occurred after drilling stopped and the visible dust dissipated, opening the door, even briefly, produced a ninefold increase in respirable dust concentrations inside the cab during many drill steel changes made over a working shift.

CONTROLLING HAULAGE ROAD DUST

Off-road haul trucks used in the mining industry typically contribute most of the dust emissions at a mine site. Although most of the airborne dust generated from unpaved haulage roads is nonrespirable, up to 20% is in the respirable size range [Organiscak and Reed 2004]. The most common method of haul road dust control is surface wetting with plain water, but others include adding hygroscopic salts, surfactants, soil cements, bitumens, and films (polymers) to the road surface [NIOSH 2003a; USBM 1987a]. Figure 5-3 shows the effectiveness of road wetting with water on airborne respirable dust generation measured next to an unpaved haul road [Organiscak and Reed 2004]. The road was wetted in the morning and dried out in the afternoon. Although the road treatment methods have been shown to be very effective, their application generally involves continual maintenance due to road degradation from traffic, dry climatic conditions, and material spillage on the road. Road dust generation can be inevitable at times during the mining operation, until controls are applied. Given their mobility, trucks have the potential for exposing downwind mine workers to respirable dust, as well as other truck drivers traveling on the haul road. NIOSH has recently studied the size characteristics, concentrations, and spatial variation of airborne dust generated along unpaved mine haulage roads to examine the potential human health and safety impacts of this dust source and is examining other avenues of truck dust mitigation. Techniques for controlling haulage road dust are summarized below.

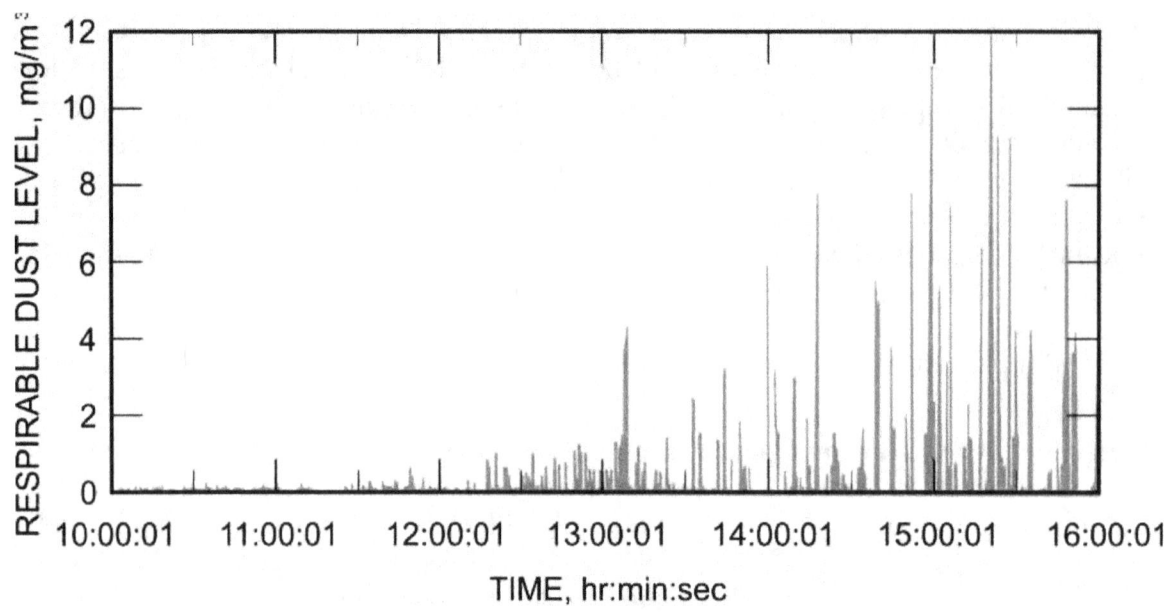

Figure 5-3. Increase in dust when a wet haul road dries.

Methods for Controlling Haulage Road Dust Exposures

- **Treatment of unpaved road surfaces.** Figure 5-3 shows the effectiveness of road wetting on respirable dust next to the road and its time frame of effectiveness [Organiscak and Reed 2004]. Other haulage road treatments, such as hygroscopic salts, surfactants, soil cements, bitumens, and films (polymers), can extend the time of effectiveness between treatments to up to several weeks [NIOSH 2003a; USBM 1987a].

- **Increase distance between vehicles traveling the haul road.** Airborne dust concentrations generated from haulage roads rapidly decreased and approached ambient air dust levels 100 feet from the road [Organiscak and Reed 2004]. This road dust dissipation and dilution occurrence provides administrative and mine planning controls to reduce worker dust exposure. The distance placed between trucks not following within 20 seconds of each other can result in a 41–52% reduction in airborne respirable dust exposure to the following truck [Reed and Organiscak 2006]. Finally, road layout and traffic patterns that can be economically incorporated into the mine plan could also isolate the haul road dust sources from other workers [Organiscak and Reed 2004].

CONTROLLING DUST AT THE PRIMARY HOPPER DUMP

Ore is normally loaded into haul trucks from the pit or quarry and driven to the primary crusher location. This ore is either dumped directly from the haul truck into the primary ore hopper feeding a crusher or dumped into a stockpile. If it is stockpiled, a front-end loader then takes the ore and dumps it into the primary hopper. In either case during this dumping process, a dust cloud is billowed out of the hopper and rolled back under the truck bed or front-end loader bucket. Dust in the ore is released from the large volume of ore product being dumped in a short period of time, which quickly displaces the air in the hopper and transports the airborne dust released from dumping. If the equipment operators dumping the ore into the hopper have an effective enclosed cab filtration system (as described earlier) their exposure to this dust would be reduced. However, if other mine personnel, such as crusher operators and/or maintenance workers, work near this primary dump, they can be exposed to this airborne dust. Several effective control methods include enclosing the hopper dump and using water sprays to suppress and contain the dust from rolling back out of the enclosure.

Key Factors for Controlling Dust from the Primary Dump

- **Enclose the primary hopper dump.** Walls can be constructed around the primary dump location to form an enclosure that must be custom designed to accommodate the dump vehicles being used. Walls can be either stationary (rigid) or movable (flexible material or curtains) based upon maintenance access within parts of the enclosure. Staging curtains (sometimes called stilling curtains) can be used in the enclosure to break up the natural tendency for dust to billow out of the primary dump hopper when a large volume of product is dumped in a very short time period (see Figure 5-4) [Weakly 2000]. Another option to restrict the dust from escaping the enclosure is using panels of flexible plastic stripping on the dump side of the enclosure. This plastic stripping employs an overlapping sequence which provides for a very effective seal and resists damage if contacted by the bucket of the front-end loader or the bed of the haul truck during dumping. Finally, a local exhaust ventilation (LEV) system can be used to filter the dust-laden air from the enclosed hopper area. This would be most appropriate when the primary dump is at a location where the dust could enter an adjoining structure or impact outside miners. Since hoppers are usually large, a significant amount of airflow would be required to create sufficient negative pressure to contain the dust cloud. This approach would be a more expensive alternative than using wet suppression [MSA Research Corp 1978].

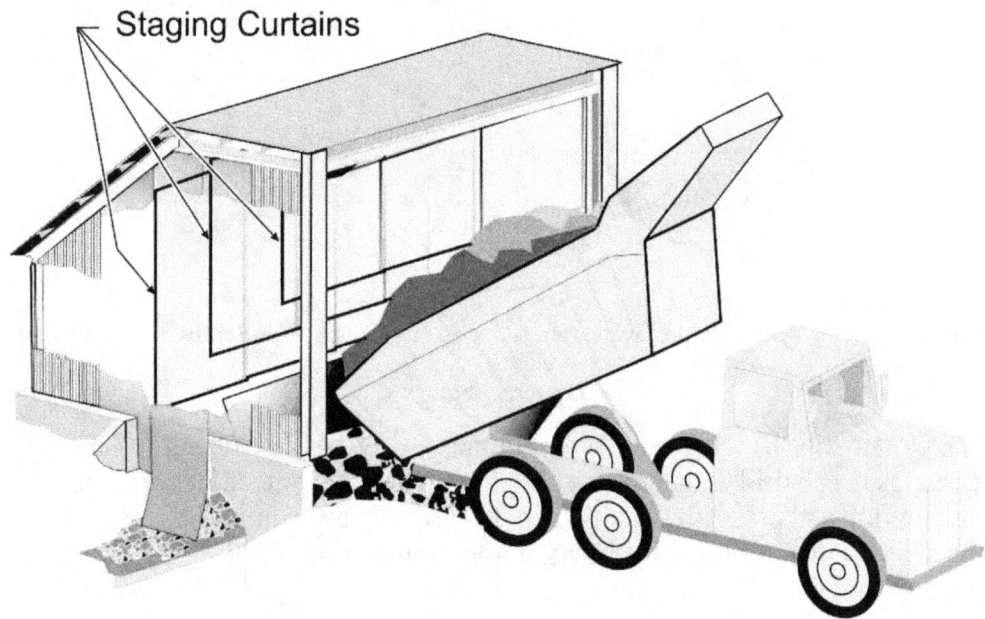

Figure 5-4. Staging curtains used to prevent dust from billowing out of enclosure.

- **Use water sprays to suppress the dust in the enclosure.** Water sprays directed at the ore dumped into the hopper will wet the material and suppress some of the generated airborne dust. A good starting point is to add 1% moisture by weight [MVS 1974]. This percentage can be adjusted based upon the improvement gained from additional moisture versus any consequences from adding too much. Since continuous use of water sprays during long periods of idle time between dumping can have adverse operational effects, the water sprays can be activated during the actual dump cycle through the use of a photo cell or a mechanical switching device. A delay timer can also be used in this application so that the sprays continue to operate and suppress dust for a short time period after the dump vehicle has moved away.

- **Prevent the dust from rolling back under the dump vehicle.** A tire-stop water spray system is recommended to reduce the dust liberated due to rollback under the dumping mechanism. A tire stop or Jersey barrier should be positioned at the most forward point of dumping for the primary hopper. A water spray system should be attached to the back side of this tire stop to knock down and force the dust that would otherwise roll back under the dumping mechanism into the hopper. Additionally, a shield should be placed over this water spray manifold to protect it from damage from falling ore (Figure 5-5). Finally, a system should also be incorporated that allows the water sprays to only be activated during the actual dumping process, as previously discussed.

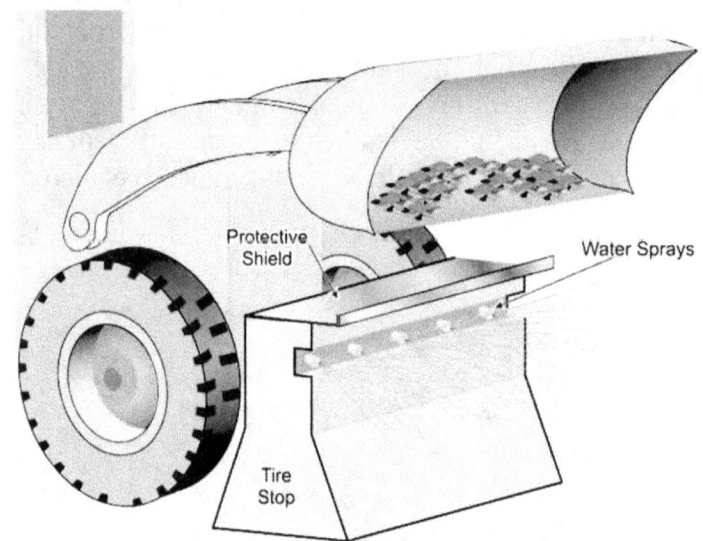

Figure 5-5. Tire-stop water spray system reduces dust rollback under the dumping vehicle.

REFERENCES

Cecala AB, Organiscak JA, Heitbrink WA, Zimmer JA, Fisher T, Gresh RE, Ashley JD [2004]. Reducing enclosed cab drill operator's respirable dust exposure at surface coal operation with a retrofitted filtration and pressurization system. In: SME Transactions 2003, Vol. 314. Littleton, CO: Society for Mining, Metallurgy and Exploration, Inc., pp. 31–36.

Cecala AB, Organiscak JA, Zimmer JA, Heitbrink WA, Moyer ES, Schmitz M, Ahrenholtz E, Coppock CC, Andrews EH [2005]. Reducing enclosed cab drill operator's respirable dust exposure with effective filtration and pressurization techniques. J Occup Environ Hyg 2:54–63.

Cecala AB, Organiscak JA, Zimmer JA, Moredock D, Hillis M [2007]. Closing the door to dust when adding drill steels. Rock Prod *October*:29–32.

CDC [2000]. Silicosis screening in surface coal miners—Pennsylvania, 1996–1997. MMWR *49*(27):612–615.

Chekan GJ, Colinet JF [2003]. Retrofit options for better dust control. Aggregates Manag *8*(9):9–12.

Listak JM, Reed WR [2007]. Water separator shows potential for reducing respirable dust generated on small-diameter rotary blasthole drills. Int J Min Reclam Environ *21*(3):160–172.

MSA Research Corp [1978]. Improved dust control at chutes, dumps, transfer points, and crushers in noncoal mining operations. NTIS no. PB 297-422. By Rodger SJ, Rankin RL, Marshall MD. Evans City PA: MSA Research Corp. U.S. Bureau of Mines contract no. H0230027.

MSHA [2009]. MSHA Standardized Information System, Arlington VA: U.S. Department of Labor, Mine Safety and Health Administration.

MVS [1974]. Sources and methods of control of dust. By Quilliam JH. In: The ventilation of South African gold mines. Yeoville, Republic of South Africa: The Mine Ventilation Society of South Africa.

NIOSH [1992]. NIOSH alert: request for assistance in preventing silicosis and deaths in rock drillers. Cincinnati, OH: U.S. Department of Health and Human Services, Centers for Disease Control, National Institute for Occupational Safety and Health, DHHS (NIOSH) Publication No. 92-107.

NIOSH [1998]. NIOSH hazard control: New shroud design controls silica dust from surface mine and construction blast hole drills. HC 27. By Page SJ, Organiscak JA, Flesch JP, Hagedorn RT. Cincinnati, OH: U.S. Department of Health and Human Services, Centers for Disease Control and Prevention, National Institute for Occupational Safety and Health, DHHS (NIOSH) Publication No. 98-150.

NIOSH [2001]. Technology News 487: Sweeping compound application reduces dust from soiled floors within enclosed operator cabs. U.S. Department of Health and Human Services, Centers for Disease Control and Prevention, National Institute for Occupational Safety and Health.

NIOSH [2003a]. Handbook for dust control in mining. By Kissell, FN. Cincinnati, OH: U.S. Department of Health and Human Services, Centers for Disease Control and Prevention, National Institute for Occupational Safety and Health, DHHS (NIOSH) Publication No. 2003-147.

NIOSH [2003b]. Work-related lung disease surveillance report 2002. Cincinnati, OH: U.S. Department of Health and Human Services, Centers for Disease Control and Prevention,

National Institute for Occupational Safety and Health, Division of Respiratory Disease Studies, DHHS (NIOSH) No. 2003-111.

NIOSH [2005]. Technology News 512: Improve drill dust collector capture through better shroud and inlet configurations. By Organiscak JA, Page SJ. U.S. Department of Health and Human Services, Centers for Disease Control and Prevention, National Institute for Occupational Safety and Health, DHHS (NIOSH) Publication No. 2006–108.

NIOSH [2007]. Technology News 528: Recirculation filter is key to improving dust control in enclosed cabs. By Organiscak JA, Cecala AB. U.S. Department of Health and Human Services, Centers for Disease Control and Prevention, National Institute for Occupational Safety and Health, DHHS (NIOSH) Publication No. 2008-100.

Organiscak JA, Page SJ [1999]. Field assessment of control techniques and long-term dust variability for surface coal mine rock drills and bulldozers. Int J Surf Min Reclam Environ *13*(4):165–172.

Organiscak JA, Reed WR [2004]. Characteristics of fugitive dust generated from unpaved mine haulage roads. International Journal of Surface Mining, Reclamation, & Environment, Vol. 18, No. 4, pp. 236–252.

Organiscak JA, Cecala AB, Thimons ED, Heitbrink WA, Schmitz M, Ahrenholtz E [2004]. NIOSH/industry collaborative efforts show improved mining equipment cab dust protection. In SME Transactions 2003, Vol. 314. Littleton, CO: Society for Mining, Metallurgy and Exploration, Inc., pp. 145–152.

Page SJ [1991]. Respirable dust control on overburden drills at surface mines. In: Proceedings of the American Mining Congress Coal Convention 1991. Washington, DC: American Mining Congress, pp. 523–539.

Page SJ, Organiscak JA [1995]. Taming the dust devil: an evaluation of improved dust controls for surface drills using rotoclone collectors. Eng Min J *November*:30–31.

Reed WR, Organiscak JA [2006]. The evaluation of dust exposure to truck drivers following the lead haul truck. In: SME Transactions 2005, Vol. 318. Littleton, CO: Society for Mining, Metallurgy and Exploration, Inc., pp. 147–153.

Reed WR, Organiscak JA, Page SJ [2004]. New approach controls dust at the collector dump point. Engin Min J *July*:29–31.

USBM [1987a]. Fugitive dust control for haulage roads and tailing basins. By Olson KS, Veith DL. Washington, DC: U.S. Department of the Interior, U.S. Bureau of Mines, RI 9069.

USBM [1987b]. Technology News 286: Optimizing dust control on surface coal mine drills. By Page SJ. Washington, DC: U.S. Department of the Interior, U.S. Bureau of Mines.

USBM [1988]. Technology News 308: Impact of drill stem water separation on dust control for surface coal mines. By Page SJ. Washington, DC: U.S. Department of the Interior, U.S. Bureau of Mines.

USBM [1995]. Technology News No. 447: Dust collector discharge shroud reduces dust exposure to drill operators at surface coal mines. By Organiscak JA, Page SJ. Washington, DC: U.S. Department of the Interior, U.S. Bureau of Mines.

Weakly A [2000]. Controlling dust without using bag houses. Coal Age *November*:24–26.

www.ingramcontent.com/pod-product-compliance
Lightning Source LLC
Chambersburg PA
CBHW081841170526
45167CB00007B/2873